Trainee Guide

Masonry

Level 1

National Center for Construction Education and Research

*Wheels of Learning
Standardized Craft Training*

Prentice Hall
Upper Saddle River, New Jersey   Columbus, Ohio

Printed in the United States of America

11

ISBN: 0-13-012268-8

Prentice-Hall International (UK) Limited, *London*
Prentice-Hall of Australia Pty. Limited, *Sydney*
Prentice-Hall of Canada, Inc., *Toronto*
Prentice-Hall Hispanoamericana, S. A., *Mexico*
Prentice-Hall of India Private Limited, *New Delhi*
Prentice-Hall of Japan, Inc., *Tokyo*
Pearson Education Asia, *Singapore*
Editora Prentice-Hall do Brasil, Ltda., *Rio de Janeiro*

Preface

This volume is one of many in the *Wheels of Learning* craft training program. This program, covering more than 20 standardized craft areas, including all major construction skills, was developed over a period of years by industry and education specialists. Sixteen of the largest construction and maintenance firms in the U.S. committed financial and human resources to the teams that wrote the curricula and planned the national accredited training process. These materials are industry-proven and consist of competency-based textbooks and instructor guides.

The *Wheels of Learning* was developed by the National Center for Construction Education and Research in response to the training needs of the construction and maintenance industries. The NCCER is a nonprofit educational entity affiliated with the University of Florida and supported by the following industry and craft associations:

Partnering Associations
- ABC Texas Gulf Coast Chapter
- American Fire Sprinkler Association
- American Society for Training and Development
- American Vocational Association
- American Welding Society
- Associated Builders and Contractors, Inc.
- Associated General Contractors of America
- Carolinas AGC, Inc.
- Carolinas Electrical Contractors Association
- Construction Industry Institute
- Design-Build Institute of America
- Merit Contractors Association of Canada
- Metal Building Manufacturers Association
- National Association of Minority Contractors
- National Association of Women in Construction
- National Insulation Association
- National Ready Mixed Concrete Association
- National Utility Contractors Association
- National Vocational Technical Honor Society
- Painting and Decorating Contractors of America
- Portland Cement Association
- Steel Erectors Association of America
- University of Florida
- Women Construction Owners and Executives, USA

Some of the features of the *Wheels of Learning* program include:
- A proven record of success over many years of use by industry companies.
- National standardization providing "portability" of learned job skills and educational credits that will be of tremendous value to trainees.
- Recognition: upon successful completion of training with an accredited sponsor, trainees receive an industry-recognized certificate and transcript from the NCCER.
- Each level meets or exceeds Bureau of Apprenticeship and Training (BAT) requirements for related classroom training (CFR 29:29).
- Well illustrated, up-to-date, and practical information. All standardized manuals are reviewed annually in a continuous improvement process.

Acknowledgments

This manual would not exist were it not for the dedication and unselfish energy of those volunteers who served on the Technical Review Committee. A sincere thanks is extended to:

Mark A. Anderson	Davis "Dick" Mitchell	Arnold Schueck
Jesse Fox	Samuel A. Nagel	Greta Soderman
Sam McGee	Robert F. Quider	Roy V. Swindal

Contents

Introduction to Masonry
Module 28101

Masonry Trainee Task Module 28101

NATIONAL
CENTER FOR
CONSTRUCTION
EDUCATION AND
RESEARCH

INTRODUCTION TO MASONRY

OBJECTIVES

Upon completion of this module, the trainee will be able to:

1. Review the history of masonry.
2. Describe modern masonry materials and methods.
3. Understand career ladders and advancement possibilities in masonry work.
4. Describe the skills, attitudes, and abilities needed to work as a mason.

Prerequisites

Successful completion of the following Task Modules is recommended before beginning study of this Task Module: Core Curricula.

Required Trainee Materials

1. Trainee Task Module
2. Pencil and paper
3. Appropriate Personal Protective Equipment

COURSE MAP

This course map shows all of the task modules in the first level of the Masonry curricula. The suggested training order begins at the bottom and proceeds up. Skill levels increase as a trainee advances on the course map. The training order may be adjusted by the local Training Program Sponsor.

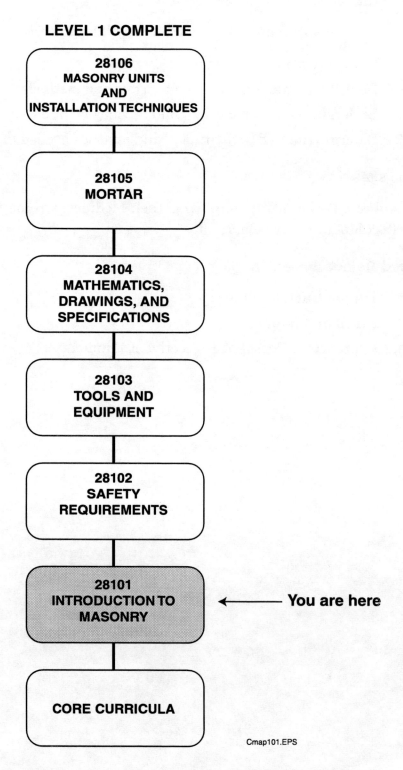

LEVEL 1 COMPLETE

28106
MASONRY UNITS
AND
INSTALLATION TECHNIQUES

28105
MORTAR

28104
MATHEMATICS,
DRAWINGS, AND
SPECIFICATIONS

28103
TOOLS AND
EQUIPMENT

28102
SAFETY
REQUIREMENTS

28101
INTRODUCTION TO
MASONRY   ← You are here

CORE CURRICULA

Cmap101.EPS

TABLE OF CONTENTS

Trade Terms Introduced In This Module

Admixture: A compound added to concrete to change its appearance, setting time, water retention, or other qualities.

Ashlar: A squared or rectangular cut stone masonry unit; or, a flat-faced surface having sawed or dressed bed and joint surfaces.

ASTM: American Society for Testing and Materials, publisher of masonry standards.

Course: A row or horizontal layer of masonry units.

Cube: A strapped bundle of approximately 500 standard bricks, or 90 standard blocks, usually palletized. The number of units in a cube will vary according to the manufacturer.

Facing: That part of a masonry unit or wall that shows, after construction; the finished side of a masonry unit.

Footing: The base for a masonry unit wall, or concrete foundation, that distributes the weight of the structural member resting on it.

Grout: A mixture of portland cement, lime, and water, with or without fine aggregate, with a high enough water content that it can be poured into spaces between masonry units and voids in a wall.

Joints: The area between each brick or block that is filled with mortar.

Mason: A person who assembles masonry units by hand, using mortar, dry stacking, or mechanical connectors.

Masonry unit: Any building block made of brick, cement, ashlar, clay, adobe, rubble, glass, tile, or any other material, that can be assembled into a structural unit.

Mortar: A mixture of portland cement, lime, fine aggregate, and water, plastic or stiff enough to hold its shape between masonry units.

Nonstructural: Not bearing weight other than its own.

PSI: Pounds per square inch. A unit of pressure or stress.

Structural: Bearing weight in addition to its own.

Tuckpointing: Filling fresh mortar into cutout or defective joints in masonry.

Weephole: A small opening in mortar joints or faces to allow the escape of moisture.

Wythe: A continuous section of masonry wall, one masonry unit in thickness, or that part of a wall which is one masonry unit in thickness.

1.0.0 INTRODUCTION

This module introduces the course in masonry. Masonry is a three-level curriculum designed to provide you with the knowledge and skills needed to work as a **mason**.

The first level introduces you to masonry history, safety, tools, materials, and techniques. This information can get you started as an apprentice mason. The second level covers elevated work, plan reading, fireplace building, and advanced techniques in many aspects of the mason's work. The third level brings more advanced techniques, plus project planning and supervision.

This first module of the first level introduces the history and the present scope of masonry, and the work of masons. The first bricks unearthed by archaeologists are about 10,000 years old. These bricks were made of hand-shaped, dried mud. The newest masonry materials include noncombustible tiling that covers the nose cones of space ships. This module will give you an idea of the distance masonry has traveled.

These definitions stretch from the historic past to the present.

Masons are persons who handle **masonry units** as they do masonry work.

Masonry units are blocks of natural or manufactured brick, cement, **ashlar**, clay, adobe, rubble, glass, tile, or any other material, that can be assembled into a **structural** unit or some part of a building.

In masonry work, masonry units are assembled by hand, using **mortar**, dry-stacking, or mechanical connectors. Despite all the advances of the last 200 years, masonry work is still hand crafted, just as it was thousands of years ago.

Common masonry elements are paving or other slabs or floors, foundations, and loadbearing or veneer walls. Uncommon masonry elements include roofs, balustrades, flying buttresses, bridges, pergolas, columns, and other architectural elements. With its versatility, modern masonry assembles into almost any kind of structural or architectural element.

2.0.0 HISTORY

Masonry is one of the oldest building materials. Its recovered history dates back to ancient times. For centuries, masonry and masons have been vital parts of the economy. During the later part of the Industrial Revolution, the prestige and usefulness of masonry construction seemed to be under a shadow. The middle of the twentieth century brought changes that increased the demand for masonry and masons. Masonry became part of the modern construction process, and the mason became an important part of the modern construction team.

2.1.0 FROM PREHISTORY TO THE INDUSTRIAL REVOLUTION

Masonry and masons have been visible in the human picture for thousands of years. The remains of stone buildings and religious monuments are among the oldest known traces of humanity. The massive fortress walls and beehive tombs at Mycene, Greece, date back to 13,000 years BCE (before the common era); the early walls of Jericho date back to 8,000 BCE. These are made of massive ashlars laid so closely that you cannot fit a piece of paper between them.

Brick is the oldest manufactured building material, invented thousands of years ago. The Hanging Gardens of Babylon hung down from brick towers. These hand-formed mud bricks were reinforced with straw and dried in the sun. They were stacked with wet mud between them. Sometimes they were covered with another coat of mud, which was decorated. This was a common and effective building technique for centuries. Some 8,000-year-old bricks (*Figure 1*) have been recovered from Jericho. The sun-dried bricks have a row of thumbprints along their tops. Later, bricks had the king's name and the date stamped on the top; this practice is very useful to today's archaeologists working to date their excavations. Today's bricks are stamped with the manufacturer's name in the same place on the top of the brick. Handmade clay brick is still in use in some parts of the Near East and Africa.

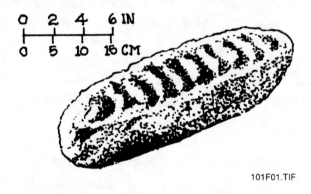

101F01.TIF

Figure 1. Hand-Formed, Sun-Dried Brick

Wooden molds for shaping clay bricks can be traced to the Tigris-Euphrates valley by 4,000 BCE. Molds, and standardization of brick sizes, became widespread during the early Bronze Age, about 3,000 BCE. Kiln-fired bricks did not appear until after 2,000 BCE, long after fired pottery was used everywhere. With kiln firing came glazing, adapted from pottery techniques. The famous lion walls of the Tigris-Euphrates valley, *Figure 2*, are made of fired and glazed bricks, sculpted into high relief before firing. The kiln-firing techniques refined in ancient India, Babylonia, and Rome are the same ones used today.

Figure 2. Lion Walls At Khorsabad, Iran

Early bricks led to early brick architecture. Someone had the idea of laying brick in different patterns instead of simply stacking them. *Figure 3* shows a herringbone pattern, seen in ancient walls still standing today. These walls had mortar holding the bricks together. The first mortar was wet mud. Along with firing and glazing brick, the Babylonians developed two new types of mortar. These were based on mixing lime or pitch (asphalt) with the mud.

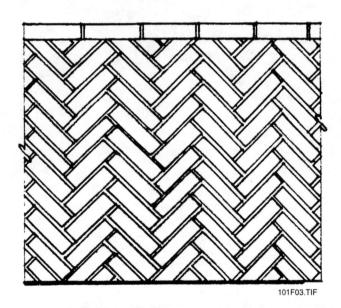

Figure 3. Herringbone Brick Wall

A later boost for brick architecture was the development of the dome and arch, *Figure 4.* With domes and arches, early masons could build larger and higher structures, with more open space inside.

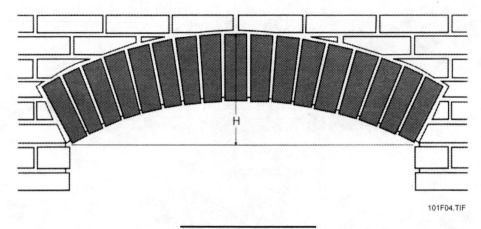

101F04.TIF

Figure 4. Radial Arch

The Romans refined arches and domes and built large-scale brickyards. They covered the Roman Empire with Roman roads that carried Roman bricks, mortar, arches (*Figure 5*), and domes, along with Roman civilization, to the known world. The Romans refined the Babylonian lime mortar by developing a form of cement that was a waterproof mortar. This mortar was useful for brick and for stone construction as well. It was also applied as a finish coat to the exterior of the surface as an early form of stucco.

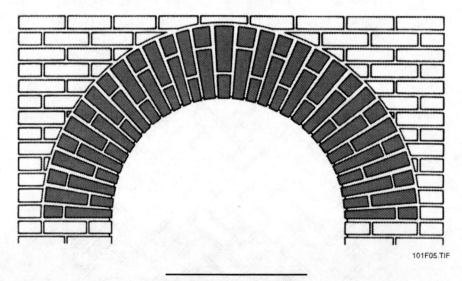

101F05.TIF

Figure 5. Roman Arch

The Romans standardized the sizes of their brick. "Roman brick" is a recognized standard size for bricks today. They produced highly ornate brick architecture, using specialized brick shapes and varied brick colors and glazes. As the Roman Empire declined, the art of intricate brickwork declined in Europe and England, but flourished in Constantinople, Russia, and the Middle East.

Without major brickwork construction, masons in western Europe earned their wages as stone workers for several centuries. Large-scale brick construction projects were started again after the Norman Conquest of 1066. The Normans built many castles in England and imported brick and the masons to lay the brick. This construction boom boosted the trade economy for northern Europe. It also boosted the status of masons. This boom also set a standard for brick; the "Norman brick" is still a recognized and widely used size of brick.

During the twelfth century CE (common era), the arch grew higher (*Figure 6*). As builders learned more about supporting weight, masonry blossomed. The architectural ideal changed from the Roman to the new Gothic style. Masons were in great demand as many stone and brick cathedrals were built across Europe.

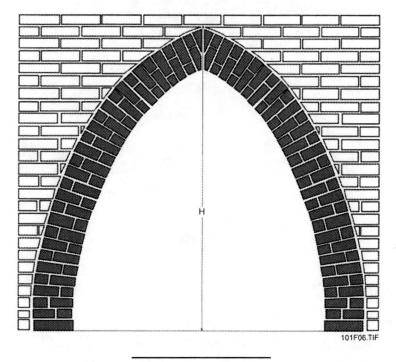

101F06.TIF

Figure 6. Gothic Arch

Figure 7 shows the west facade of Reims Cathedral, one of the most beautiful examples of Gothic masonry work in the world. This cathedral contains many Gothic arches, as can be seen in the lower, middle, and upper sections of the structure.

Masons were among the first to organize into craft guilds during the Middle Ages. By then, the skills of the master masons had expanded. They did all the tasks that the architect, engineer, general contractor, mason contractor, and materials supplier do today.

Figure 7. Reims Cathedral

2.2.0 THE RISE OF MASONS

As the demand for more elaborate construction grew, the need for skilled workers became greater. The earliest definition of "mason" was "a builder and worker in stone; a worker who dresses and lays stone." (Layamon, *Brut*, 1205 CE.) By the middle 1300s, masons had organized into early unions, or guilds, across most of Europe.

Guilds controlled the practice of the craft by monitoring the skill level of the craftworker. Early masons' guilds (Condor, *Hole Craft*, 1376) recognized three levels of work.

- The ligier or rough mason was the equivalent of the apprentice of today. The ligier worked where the guild would allow, under supervision of masters or journeymen. Ligiers served an apprenticeship of three to five years. They were examined by the guild before being allowed to work independently as a mason.

- The journeyman, or mason, was a skilled worker allowed to work on finer jobs without supervision. Journeymen were also free to move to wherever the work was. The local guild set the time, as long as 10 years, for this stage. At the end, the journeyman could take an examination before the guild and be awarded the status of master. Not all masons reached master status.

- The master mason ran the business, designed structures, employed journeymen, and trained apprentice masons. The master also held office, served on boards, and ran the affairs of the masons' guild.

The local guilds continued to control the practice of the mason's craft for centuries. They monitored training, judged disputes, and shared knowledge between members. They collected dues, provided some support to widows and the ill, and celebrated special mason's holidays.

Nonmembers were not allowed to know the secrets of the guilds. Guilds recognized or did not recognize masons certified by other local guilds. Secret signs and passwords were developed so that masons with high skills could recognize each other in different places.

In the early 1700s, Masonic Temples and Orders of Freemasonry were founded in Europe. These organizations were political and spiritual but based on many of the ideas of the masons' guilds. The pyramid seal on the back of the American dollar bill is an isolated legacy of the masons' guilds.

The strong craftworker tradition, the apprenticeship programs, and the skills and techniques of the work itself are direct legacies of the masons' guilds.

2.3.0   AFTER THE INDUSTRIAL REVOLUTION

Masonry came to the New World with the earliest European settlers. Jamestown, the site of the James River Colony, had its masons by 1611. They made brick and built the first fired brick buildings in America. *Figure 8* shows a building in the style popular then. Jamestown became the home of two brickyards, or brick factories, which operated profitably for the next 50 years. On the west coast, the Spaniards built missions and settlements of adobe throughout California, New Mexico, and Arizona. Masonry continued to be the premier building material, in the New World and the Old, until the Industrial Revolution.

101F08.TIF

Figure 8. Typical Brick Outbuilding of the Colonial Period

The first reinforced stone and concrete building, the Eddystone Lighthouse, was built in England in 1774. *Figure 9* shows the ruins of this structure, and its replacement built in 1884. The newer tower benefited from the invention of portland cement in 1824. It incorporated improvements in iron and steel manufacturing as well.

101F09.TIF

Figure 9. Eddystone Lighthouse

These changes in building materials brought changes in building techniques. The interests of engineers and craftworkers turned to the new building materials. Engineers developed a science of building tall by reinforcing concrete with iron and steel. During this time, the popularity of masonry was replaced by these new materials.

The last all-masonry, unreinforced brick high-rise building was the 16-story, block-long Monadnock Building. Shown in *Figure 10*, it was built in Chicago in 1891. Its walls tapered up to 6 feet thick at the first floor. The thick walls were considered necessary to support the weight of the structure. This technique was very uneconomical compared to the costs of building with the newer materials. The architect, Daniel Root, suggested an iron framework for the building. The owners decided they did not trust it. The iron might rust away, but the brick would stand. The Monadnock Building, still in use, represents the last monument of the era before modern masonry.

By 1895, masonry techniques did not seem suited to modern high-rise construction. The new tall buildings generated heavy loads that required calculation of precise support. Unfortunately, rapid advances in concrete and steel engineering and design were not matched by advances in masonry engineering. Masonry was no longer the first and the best building material.

Figure 10. Monadnock Building, Chicago

It was not until the 1920s, in India, that research began into reinforcing masonry. This led to new systems of low-cost construction and the first understanding of masonry as a modern structural element. By the late 1940s, European engineers began studies of masonry bearing walls, developing formulas and designs. Extensive testing produced the first reliable mathematical analyses of a very old material and brought it into the new age. Six-foot walls were no longer necessary to support a high-rise masonry structure. The support and strength of reinforcement became available for masonry as well as for concrete.

Materials came into the new age also. Concrete block first appeared after 1880, and by 1917 cinderblock was patented. This process used cinders to make concrete block lightweight and fireproof. This low-cost material combined with the new building techniques to bring masonry back as a cost-effective way of building.

Masonry design and construction became economically important again in the U.S. in the 1950s. The first "engineered masonry" building code was published in 1966. Continued research over the next 20 years brought about refinements in testing and design. Testing has identified the properties of masonry in fire control, sound absorption, and thermal resistance, as well as load bearing. Engineered masonry structural systems are now part of all the major building codes in the U.S. As an example, *Table 1* shows allowable compressive stresses in masonry walls in pounds per square inch (**psi**), from the 1998 North Carolina Building Code.

ALLOWABLE COMPRESSIVE STRESSES FOR EMPIRICAL DESIGN OF MASONRY

CONSTRUCTION: COMPRESSIVE STRENGTH OF UNIT, GROSS AREA	ALLOWABLE COMPRESSIVE STRESSES GROSS CROSS SECTION AREA*	
	Type M or S Mortar	Type N Mortar
Solid masonry brick and other solid units of clay or shale; sand-lime brick or concrete brick:		
8,000+ psi	350	300
4,500 psi	225	200
2,500 psi	160	140
1,500 psi	115	100
Grouted masonry, of clay or shale: sand-lime brick or concrete brick:		
4,500+ psi	225	200
2,500 psi	160	140
1,500 psi	115	100
Solid masonry of solid concrete masonry units:		
3,000+ psi	225	200
2,000 psi	160	140
1,200 psi	115	100
Masonry of hollow loadbearing units:		
2,000+ psi	140	120
1,500 psi	115	100
1,000 psi	75	70
700 psi	60	55
Hollow walls (cavity or masonry bonded) solid units:		
2,500+ psi	160	140
1,500 psi	115	100
Hollow units	75	70
Stone ashlar masonry:		
Granite	720	640
Limestone or marble	450	400
Sandstone or cast stone	360	320
Rubble stone masonry:		
Course, rough, or random	120	100

* Gross cross-sectional area should be calculated as the actual rather than the nominal dimensions.

101T1.EPS

Table 1. Compressive Stresses For Masonry Walls

3.0.0 MASONRY TODAY

Today, masonry is used in residential, industrial, and commercial projects in patios, and in high-rise buildings. Masons belong to a union or trade association, or an informal social group, instead of a guild. They are still recognized as premier craftworkers at any construction site. Their work takes advantage of twentieth-century technology. The next sections discuss modern materials and structures.

The two main types of masonry units manufactured today are made of clay or concrete. Clay products are commonly known as brick and tile, and concrete products are commonly known as block or concrete masonry units (CMU).

3.1.0 CLAY PRODUCTS

Brick has been developed and improved upon for centuries. The modern age in brick manufacture started with the first brickmaking machine. It was powered by a steam engine and patented in 1800. The process, shown in *Figure 11*, has not changed much. The clay is mined, pulverized, and screened. It is mixed with water, formed, and cut into shape. Some plants extrude the clay, punch holes into it, then cut it into shape. Any coating or glazing is applied before the units are air dried. After drying, the brick is fired in a kiln. Because of small variations in materials and firing temperatures, not all bricks are exactly alike. Even bricks made and fired in the same batch have variations in color and shading. The brick is slowly cooled to prevent cracking. It is then bundled into **cubes** and shipped. A cube traditionally holds 500 standard bricks, or 90 blocks, although manufacturers today make cubes of varying sizes.

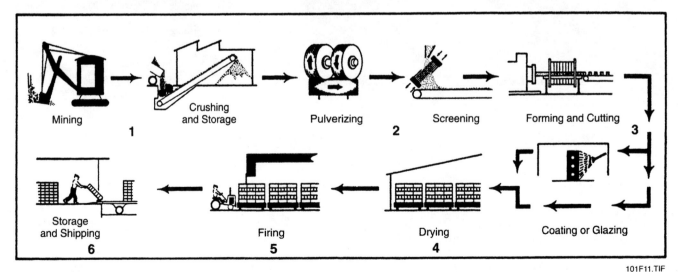

101F11.TIF

Figure 11. Brickmaking Process

Today there are over 100 commonly manufactured structural clay products. The American Society for Testing and Materials (**ASTM**) has published consensus standards for masonry design and construction. The standards cover performance specifications for manufactured masonry units. The ASTM has also specified standard sizes for various kinds of brick. *Figure 12* shows the standard sizes for today's most commonly used brick. The first six types are the most widely used. Note that the Norman brick which was mentioned earlier is among this group.

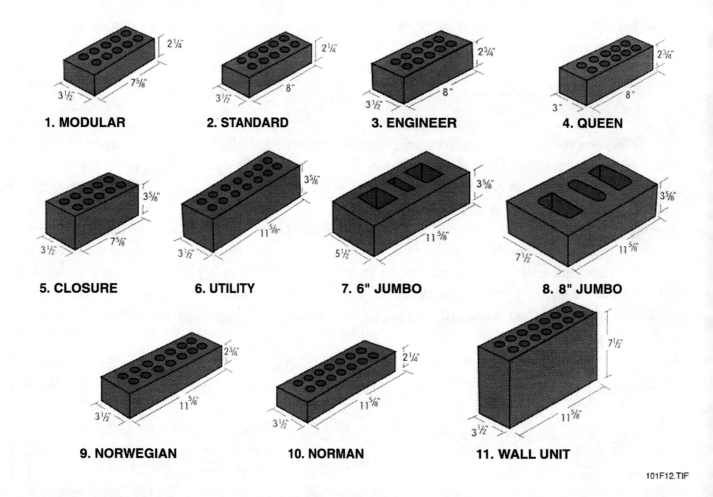

Figure 12. Standard Brick And Sizes

Structural clay products include:

- Solid masonry units, or brick
- Hollow masonry units, or tile
- Architectural terra cotta units

The next sections provide more information about these products.

3.1.1 Structural Clay/Brick

Brick is classified as solid if 25 percent or less of its surface is a hole (void). Brick is further classified by type, as building, **facing**, hollow, paving, ceramic glazed, thin veneer, sewer, and manhole. ASTM standards exist for all these types of brick. Fire brick has its own standard but is not considered a major type classification.

Brick comes in modular and nonmodular sizes, in colors determined by the minerals in the clay or by additives, in a variety of face textures, and in a rainbow of glazes. The variety is dazzling. Brick can be laid in structural bonds to create patterns in the face of a wall or walkway. *Figure 13* shows several examples of commonly used traditional bond patterns. Some bond patterns are traditional in some parts of the country. The herringbone pattern shown in *Figure 3* is still popular for walkways.

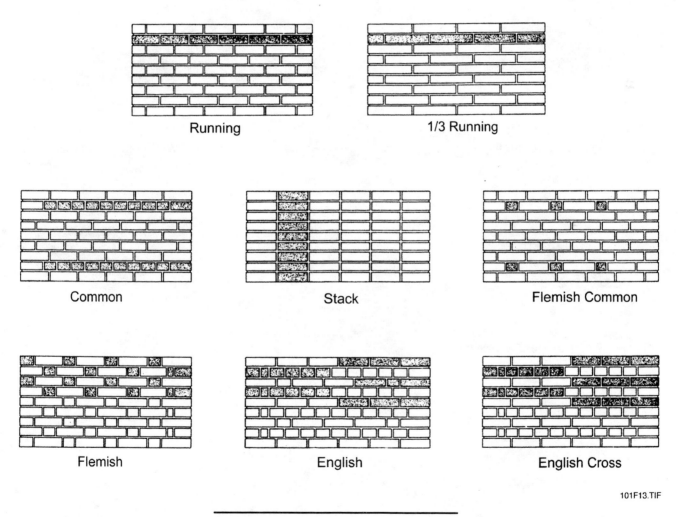

Running

1/3 Running

Common

Stack

Flemish Common

Flemish

English

English Cross

101F13.TIF

Figure 13. Traditional Masonry Bonds

Brick is also made in special shapes to form arches, sills, copings, columns, and stair treads. Custom shapes can be made to order for architectural use or for artistic use. Sculptural brick, used for the lions in *Figure 2*, is available as a custom order today. *Figure 14* shows some commonly manufactured special shapes of brick.

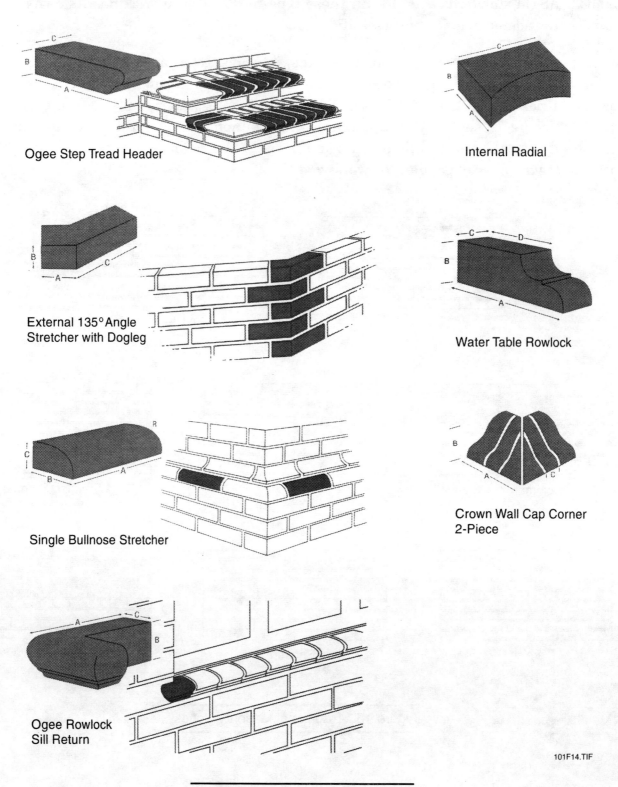

Ogee Step Tread Header

Internal Radial

External 135° Angle
Stretcher with Dogleg

Water Table Rowlock

Single Bullnose Stretcher

Crown Wall Cap Corner
2-Piece

Ogee Rowlock
Sill Return

101F14.TIF

Figure 14. Special Shapes Of Brick

3.1.2 Hollow Masonry Units/Tiles

Hollow masonry units are machine-made clay tiles extruded through a die and cut to the desired size. Hollow masonry units have less than 75 percent of their surface area solid. Hollow units are classified as either structural clay tile or structural clay facing tile.

Structural clay tile comes in many shapes, sizes, and color variations. Structural clay tile is divided into loadbearing and nonloadbearing types. *Figure 15* shows common tile shapes. Note that structural tile can be used for load bearing on its side or on its end. In some applications, structural tile is used as a backing **wythe** behind brick. Nonloadbearing tile is designed for use as fireproofing, furring, or ventilating partitions.

Structural clay facing tile comes in modular sizes as either glazed or unglazed tile. It is designed for interior applications where precise tolerances are required. A special application of clay facing tile is as an acoustic barrier. The acoustic tiles have a holed face surface to absorb sound. Clay facing tile can also be patterned by shaping the surface face.

3.1.3 Architectural Terra Cotta

Architectural terra cotta is a made-to-order product with an unlimited color range. High-temperature fired ceramic glazes are available in an unlimited color range and unlimited arrangements of parts, shapes, and sizes.

Architectural terra cotta is classified into anchored ceramic veneer, adhesion ceramic veneer, and ornamental or sculptured terra cotta. Anchored ceramic veneer is thicker than 1 inch held in place by **grout** and wire anchors. Adhesion ceramic veneer is 1 inch or thinner, held in place by mortar. Ornamental terra cotta is frequently used for cornices and column capitals on large buildings.

3.2.0 CONCRETE PRODUCTS

The history of concrete masonry units is not as long as that of brick, only a little more than a century. Concrete building blocks were invented as a result of two trends in improving masonry. One trend had to do with techniques, the other with materials. In 1850, Joseph Gibbs was trying to develop a better way to build masonry cavity walls.

Masonry is not waterproof, only water resistant. Thick walls slow down moisture so that it does not reach the inside of the wall, but thick walls are expensive to build. Cavity walls were invented to handle this problem. Over the years, many types of cavity walls have been designed to slow down or prevent moisture from reaching the inside surfaces. Today, several different designs are applied based on the availability of local materials and environmental requirements. Most of them use either brick or a combination of brick and block.

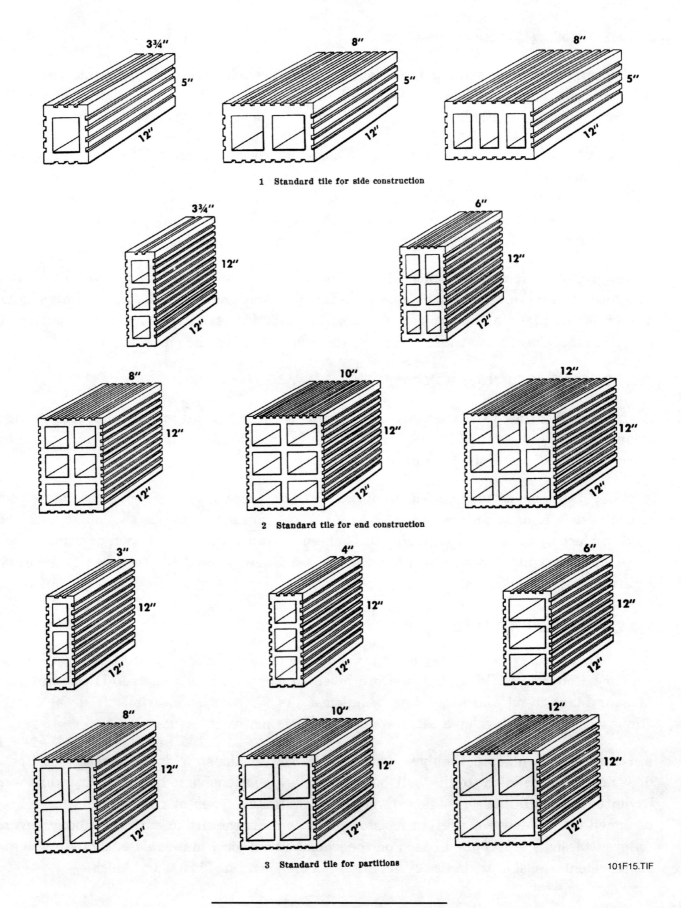

1 Standard tile for side construction

2 Standard tile for end construction

3 Standard tile for partitions

101F15.TIF

Figure 15. Structural Clay Tile Shapes

Figure 16 shows examples of both the old and newer style cavity walls. Cavity walls are made of two **courses** of masonry units, with a 2- to 4-inch gap between them. Water can get through the outside wall and run down inside the cavity without wetting the inside wall.

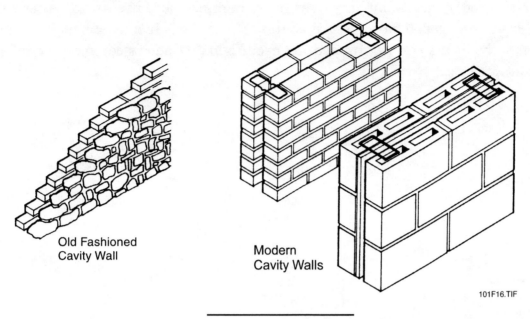

Old Fashioned
Cavity Wall

Modern
Cavity Walls

101F16.TIF

Figure 16. Cavity Walls

In his search for a faster way to build a cavity wall, Gibbs developed a block with air cells in it. This idea was refined and patented by several other people. In 1882, someone took advantage of new materials developments to make a hollow block of portland cement.

In 1900, Harmon Palmer patented a machine that made hollow concrete block. The blocks were 30 inches long, 10 inches high, 8 inches wide, heavy, and very hard to lift. Suddenly cavity walls became cheaper and faster to build. Other machines were developed, producing smaller blocks that were easier to handle but still very heavy for their size.

In 1917, Francis Straub patented a block made with the cinders left after burning coal. He used the cinders to replace the sand and small aggregates in the concrete mix. This new cinderblock was lighter, cheaper, and easier to handle. Straub's block made it possible to build a one-course wall with a built-in cavity very quickly and inexpensively. Faster machinery was developed to keep up with the demand for this new masonry material.

The demand for block increased with the rise of engineered masonry in the U.S. The production of concrete block passed that of clay brick in the 1950s. Since the 1970s, there have been more walls built of concrete block in the U.S. than those of clay brick and all other masonry materials together.

Blocks are made of water added to portland cement, aggregates (sand and gravel), and **admixtures**. The cinders have been replaced today by fly ash and other lightweight aggregates. Admixtures affect the color and other properties of the cement, such as freeze resistance, weight, and speed of setting. It is machine-molded into shape. It is compacted in the molds and cured, typically, by low-pressure or high-pressure live steam. After curing, the blocks are dried and aged. The moisture content is checked. It must be at least a specified amount before the blocks can be shipped for use. *Figure 17* shows commonly used sizes and shapes of concrete block.

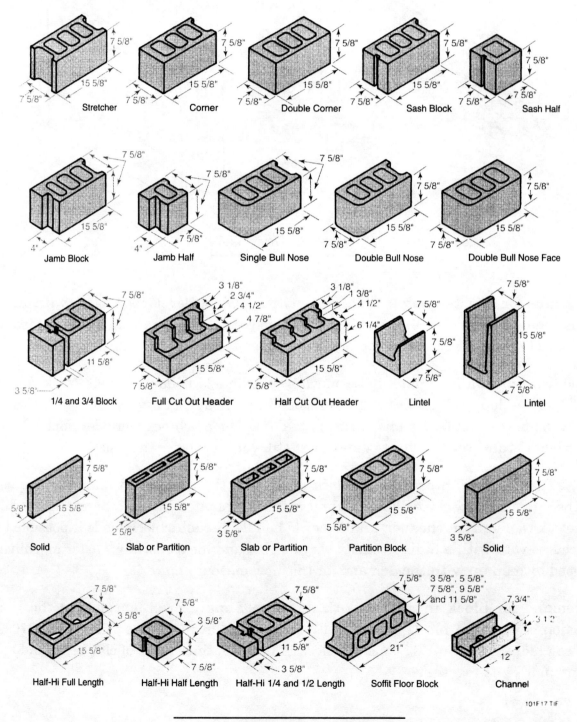

Figure 17. Common Concrete Block

Concrete masonry units are not all blocks. Concrete units fall into classifications based on intended use, size, and appearance. ASTM standards exist for:

- Loadbearing and nonloadbearing concrete block
- Concrete brick
- Calcium silicate face brick
- Prefaced or prefinished facing units
- Manholes and catch basin units

3.2.1 Block

Concrete block is a large unit, typically 8 × 8 × 16 inches, with a hollow core. Blocks come in modular sizes, in colors determined by the cement ingredients, the aggregates, or any admixtures. A variety of surface and mixing treatments can give block varied and attractive surfaces. Newer finishing techniques give block the appearance of brick, rough stone, or cut stone. Just as with clay masonry units, block can be laid in structural pattern bonds.

Figure 18 shows the names of the parts of block. Block takes less time to lay than brick. Block takes up more space, so fewer are needed. Block bed **joints** usually need mortar only on the shells and webs, so there is less mortaring as well.

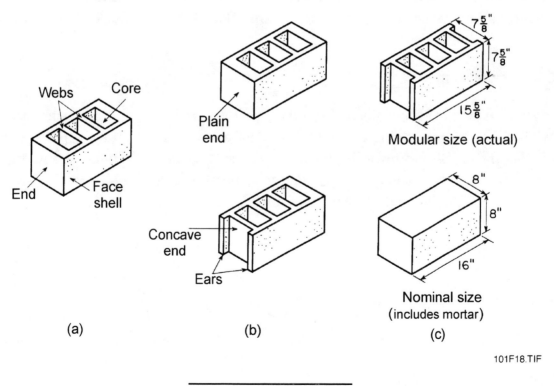

Figure 18. Parts Of Block

Concrete block comes in three weights; normal, lightweight, and aerated. Lightweight block is made with fly ash, pumice, scoria, or other lightweight aggregate. Loadbearing and appearance qualities of the first two weights are similar; the major difference is that lightweight block is easier and faster to lay. Normal weight block can be made of concrete with regular, high, and extra-high strengths. The last two are made with different aggregates and curing times. They are used to limit wall thickness in buildings over ten floors high. Aerated block is made with an admixture that generates gas bubbles inside the concrete for a lighter block.

Concrete blocks are classified as hollow or solid. As with clay products, a hollow unit has less than 75 percent of its surface area solid. Common hollow units have two or three cores. The hollow cores make it easy to reinforce concrete block walls. Grout, or steel rods and grout, can easily fill the hollow cores. Reinforcement increases loadbearing strength, rigidity, and wind resistance. Solid units have 25 percent or less of their surface hollow. Solid normal and lightweight units are for special needs, such as structures with unusually high loads, drainage catch basins, manholes, or firewalls. Aerated block is made in an oversize solid unit used for buildings.

Loadbearing block is used as backing for veneer walls, for bearing walls, and for all structural uses. Both regular and specially shaped blocks are used for paving, retaining walls, and slope protection. **Nonstructural** block is used for screening, partition walls, and as a veneer wall for wood, steel, or other backing. Both kinds of blocks come in a variety of shapes and modular sizes.

3.2.2 Concrete Brick

The length and height dimensions of regular concrete brick are the same as standard clay brick with the thickness being an additional ⅛ of an inch. A popular type is slump brick, shown in *Figure 19*. Slump brick is made from very wet concrete. When the mold is removed, the brick bulges because it is not dry enough to completely hold its shape. Slump brick looks like ashlar and adds a decorative element to a wall.

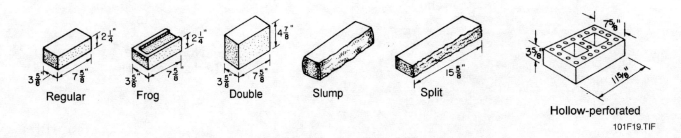

101F19.TIF

Figure 19. Slump And Other Concrete Bricks

Concrete brick is produced in a wide range of textures and finishes. It is available in specialized shapes for copings, sills, and stairs, just as clay brick is. Concrete brick is more popular in some areas of the country because it is less expensive.

3.2.3 Other Concrete Units

Concrete prefaced or precoated units are coated with colors, patterns, and textures on one or two face shells. The facings are made of resins, portland cement, ceramic glazes, porcelainized glazes, or mineral glazes. The slick facing is easily cleaned. These units are popular for use in gyms, hospital or school halls, swimming pools, and food processing plants. They come in a variety of sizes and special-purpose shapes, such as coving and bullnose corners. *Figure 20* shows commonly used concrete prefaced units.

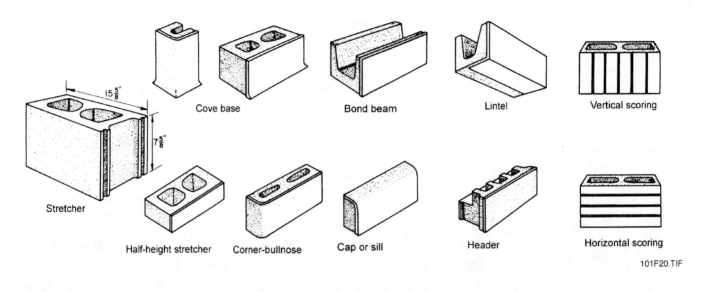

101F20.TIF

Figure 20. Prefaced Concrete Units

Concrete manhole and catch basin units are specially made with high-strength aggregates. They must resist the internal pressure generated by the liquid in the completed compartment. *Figure 21* shows the shaped units manufactured for the top of a catchment vault. These blocks are engineered to fit the vault shape and cast to specification. They are made with interlocking ends for further strength.

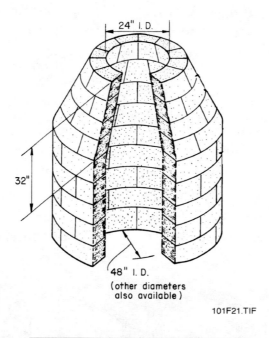

24" I.D.

32"

48" I.D.
(other diameters
also available)

101F21.TIF

Figure 21. Manhole And Vault Units

3.3.0 STONE

Stone is still used for building cathedrals, monuments, and buildings of special significance. *Figure 22* shows part of the Washington National Cathedral, under construction since 1906 and likely to be under construction for the next 50 years. The tower steeple, right, is covered with scaffolding. Note the very fine ribbed Georgian brick chimney on the vestry building in the foreground. Today, rubble and ashlar are used for dry stone walls, mortared stone walls, retaining walls, facing walls, slope protection, paving, fireplaces, patios, and walkways. Stone is also used as a veneer over brick or block. The brownstone buildings in New York and the grey stone buildings of Paris are veneer over brick. Many of the government buildings and monuments in Washington, D.C., are of stone veneer construction.

Rubble stone is irregular in size and shape. Stones collected in a field are rubble. Rubble from quarries is left where shaped blocks have been removed. It is also irregular with sharp edges. Rubble can be roughly squared with a brick hammer to make it fit together more easily.

Ashlar stone has been cut at the quarry. It has smooth bedding surfaces that stack easily. Ashlar is usually granite, limestone, marble, sandstone, or slate. Additional stone may be common in different parts of the country.

Flagstone is designed to be used for paving or floors. It is 2 inches or less thick and cut into flat slabs. Flagstone is usually quarried slate, although other stone may be popular in different areas of the country.

Figure 22. Washington National Cathedral

Stone, including flagstone, can be laid in a variety of decorative patterns. *Figure 23* shows coursed and uncoursed patterns. Even with uncoursed stone, there are suggestions of patterning. The eye of the mason generates the pattern first.

Concrete masonry units are made in shapes and colorings to mimic every kind of ashlar. These units are called cast stone and are more regular in shape and finish than natural stone. Cast stone has replaced natural stone in many commercial projects because of economy. ASTM specifications cover cast stone and natural stone.

3.4.0   MORTARS AND GROUTS

Mortar is no longer made of mud, but sometimes it is still called "mud." The first mortar was wet mud. This is still in use in some areas of the world. By the Roman period, sand was a common additive. Burnt limestone, or quicklime, was added as an ingredient around the first century BCE. Experimenting with waterproofing mortar, the Romans added volcanic ash and clay. The resulting "cement" made a strong, waterproof mortar. This made it possible to construct aqueducts and water tanks, water channels, and baths which are still in use today. Unfortunately, the Roman formula was lost over time.

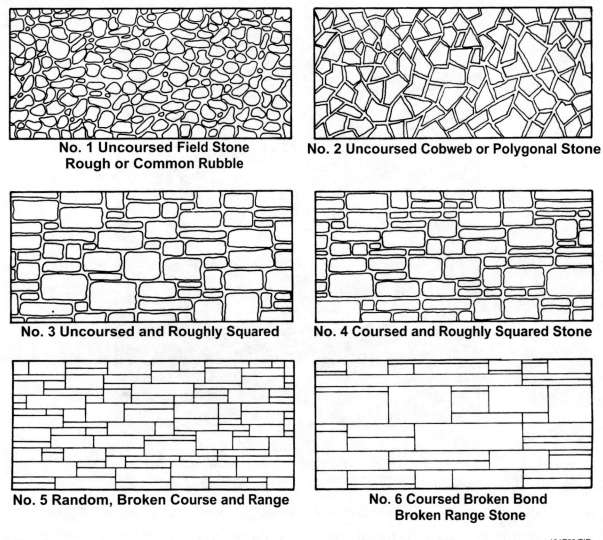

No. 1 Uncoursed Field Stone
Rough or Common Rubble

No. 2 Uncoursed Cobweb or Polygonal Stone

No. 3 Uncoursed and Roughly Squared

No. 4 Coursed and Roughly Squared Stone

No. 5 Random, Broken Course and Range

No. 6 Coursed Broken Bond
Broken Range Stone

101F23.TIF

Figure 23. Stone Wall Patterns

In 1824, portland cement was patented by Joseph Aspdin, a mason. He was trying to recreate the waterproof mortar of the Romans. By 1880, portland cement had become the major ingredient in mortar. The new, waterproof portland cement mortar began to replace the old lime and sand mixture.

Portland cement is made of ground earths and rocks burnt in a kiln to leave clinkers. The clinkers are ground to become the cement powder. Mixed with water, lime, and rocks, the cement becomes concrete. Mortar is somewhat different from concrete in consistency and use. The components and performance specifications are different also.

Modern mortar is mixed from portland cement or other cementitious material, lime, water, sand, and admixtures. The proportions of these elements determine the characteristics of the mortar. *Table 2* shows specifications for mortar strength, water retention, and air content.

**Property Specifications for
Laboratory-Prepared Mortar***

Mortar type	Minimum compressive strength, psi At 28 days	Minimum water retention, %	Maximum air content, %**
M	2500	75	12***
S	1800	75	12***
N	750	75	14***
O	350	75	14***

* Adapted from ASTM C270.

** Cement-lime mortar only (except where noted)

*** When structural reinforcement is incorporated in cement-lime or masonry
cement mortar, the maximum air content shall be 12% or 18%, respectively.

Note: The total aggregate shall be equal to not less than 2 1/4 and not more
than 3 1/2 times the sum of the volumes of the cement and lime used.

101T2.EPS

Table 2. ASTM Mortar Specifications

The two main types of mortar are as follows:

• Portland cement-hydrated lime mortars are made of portland cement, hydrated lime,
sand, and water. These ingredients are mixed at the job site by the mason.

• Masonry cement mortars are premixed with additives. The mason only adds sand and
water. The additives affect flexibility, drying time, and other properties.

Mortar is mixed to meet four sets of performance specifications as listed in the chart in
Table 2.

• *Type M* mortar, with high compressive strength, is typically used in contact with earth for
foundations, sewers, and walks. This varies with geographic location.

• *Type S* mortar, with medium strength, high bonding, and flex, is used for reinforced
masonry and veneer walls.

• *Type N* mortar, with heavy weather resistance, is used in chimneys, parapets, and
exterior walls.

• *Type O* mortar, with low strength, is used in nonloadbearing applications. It is not
recommended for professional use.

Another type of mortar, Type K, has no cement materials but only lime, sand, and water. This
type of mortar is used for preservation or restoration of historic buildings.

Grout is a mixture of cement and water, with or without fine aggregate. Wet enough to be pumped or poured, it is used in reinforcement to bond masonry and steel together. It gives added strength when it fills the cores of block walls.

4.0.0 MODERN CONSTRUCTION TECHNIQUES

This section introduces modern structures and modern construction techniques. There are several types of structures which the masonry trainee must learn to build. These structures and the techniques used to build them are basic to the craft.

4.1.0 WALL STRUCTURES

Masonry structures today take many forms in residential, commercial, and industrial construction. Modern engineering has added loadbearing strength so masonry can carry great weight without bulk. In addition to load bearing, masonry offers these advantages:

- Durability
- Ease of maintenance
- Design flexibility
- Attractive appearance
- Weather and moisture resistance
- Competitive cost

Modern engineering and ASTM standards have been applied directly to everyday masonry work. There are six common classifications of structural wall built with masonry. *Figure 24* shows some of these walls. Masonry walls can fit into more than one classification.

- Solid walls (*Figure 24a*) are built of solid masonry units with full mortar joints. Solid units have voids of less than 25 percent of their surface. These walls can have 1 or 2 wythes tied together with mortar, both loadbearing.

- Hollow walls (*Figure 24b*) are solid walls built of masonry units with more than 25 percent of their surface hollow. These can also have 1 or 2 wythes tied together with mortar, both loadbearing.

- Cavity walls (*Figure 24c*) have 2 wythes with a 2- to 4½-inch space between them. Sometimes insulation is put in the cavity. The wythes are tied together with metal ties, with both wythes loadbearing.

- Veneer walls (*Figure 24d*) are not loadbearing. A masonry veneer is usually built 1 to 2 inches away from a loadbearing stud wall or block wall. Veneer walls are used in high-rise and residential construction.

- Composite walls have different materials in the facing (outer) and backing (inner) wythes. The wythes are set with a 1-inch air space between them and are tied together by metal ties. Unlike a veneer wall, both wythes of a composite wall are loadbearing.

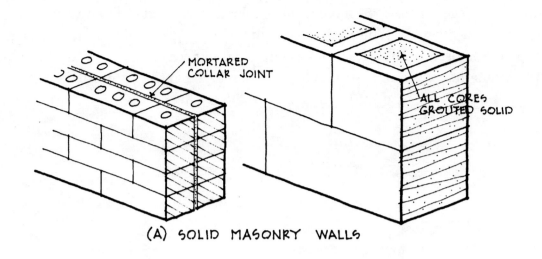

(A) SOLID MASONRY WALLS

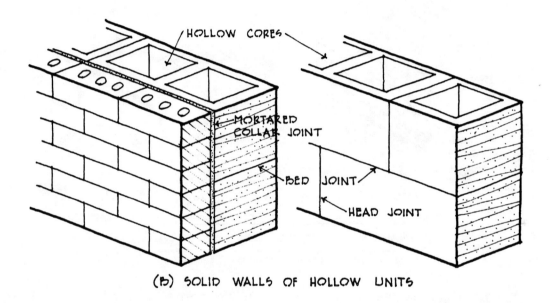

(B) SOLID WALLS OF HOLLOW UNITS

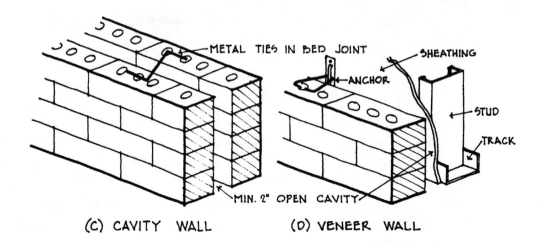

(C) CAVITY WALL   (D) VENEER WALL

101F24.TIF

Figure 24. Types Of Masonry Construction

• Reinforced walls (*Figure 25*) have steel reinforcing embedded in the cores of block units or between two wythes. The steel is surrounded with grout to hold it in place. This very strong wall is used in high-rise construction and in areas subject to earthquake and high winds. Sometimes, grout is used alone for reinforcement. The grout is pumped into the cores of the blocks or into the cavity between the wythes.

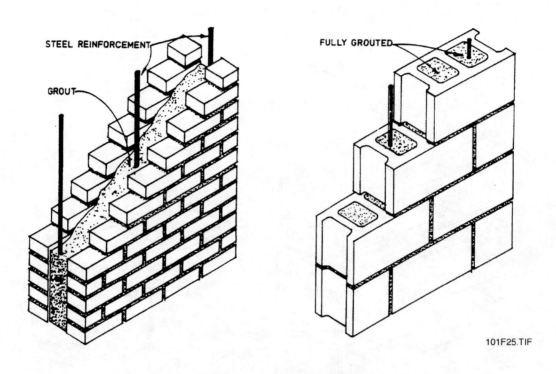

Figure 25. Reinforced Block And Brick Walls

Contemporary masonry systems are designed not as barriers to water but as drainage walls. Penetrated moisture is collected on flashing and expelled through **weepholes**. Design, workmanship, and materials are all important to the performance of masonry drainage walls.

4.2.0 MODERN TECHNIQUES

Masons today work surrounded by highly specialized hand and power tools. As you will learn in the module on tools and equipment, there are many kinds of specialized trowels (*Figure 26*) and at least six kinds each of hammers, chisels, and steel joint finishing tools. There are seven kinds of measuring and leveling tools. Power tools include several kinds each of saws, grinders, splitters, and drivers. Mortar can be mixed by hand, using special equipment, or in a power mixer. Cranes, hoists, and lifts bring the masonry units to the masons working on one of four types of steel scaffolding. The job of the mason has been made easier by **tools and equipment** to serve the mason's every need.

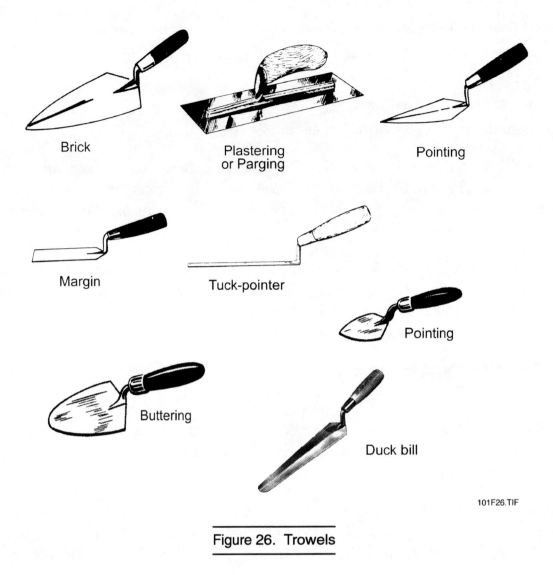

Brick

Plastering or Parging

Pointing

Margin

Tuck-pointer

Pointing

Buttering

Duck bill

101F26.TIF

Figure 26. Trowels

The only thing that has not changed is the relation between the mason and the masonry unit. The mason uses this twentieth-century wealth of tools and equipment to:

- Calculate the number and type of units needed to build a structure
- Estimate the amount of mortar needed
- Assemble the units near the work station
- Lay out the wall or other architectural structure
- Cut units to fit, as needed
- Mix the appropriate type and amount of mortar
- Place a bed of mortar on the **footing**
- Butter the head joints and place masonry units on the bed mortar
- Check that each unit is level and true
- Lay courses in the chosen bond pattern or create a new pattern
- Install ties as required for loadbearing

- Install flashing and leave weepholes as required for moisture control
- Clean excess mortar off the units as the work continues
- Finish the joints with jointing tools
- Give the structure a final cleaning
- Complete the work to specification, on time

Masonry work is still very much a craft. The relation between the mason and the masonry unit is personal. The straightness and the levelness of each masonry unit in a structure—brick, block, or rock—depend on the hand and the eye of the mason. These things have not changed in 10,000 years.

The tradition of masonry calls for a bit of art, too. The eye of the mason gets trained by work to see the subtle shadings and gradations of color. The mason learns to create a pattern and to select the right unit to complete the pattern or the shading. The mason grows skilled in building something that looks handsome and endures.

5.0.0 MASONRY AS A CAREER

Masonry offers a rewarding career for people who want to work with their hands. As masons, they will be skilled workers. They will understand the principles and practices of masonry construction. They will earn good pay and be rewarded for initiative. They will have opportunity for advancement.

Masons will continue to play an important part in building homes, schools, offices, and commercial structures. They can add artistic elements to their work and create beauty. They can be proud of their skills and the fact that they produce something people need.

Masons work on different projects, so each job is different and never boring. If they like to travel, masons can find good jobs all over the country. Masons can be independent and creative while working outdoors. Masons will always be in demand as long as construction continues.

Masons can find work on large construction projects for commercial or high-rise buildings. They can find work on small projects for building homes, patios, sidewalks, or walls. They can specialize in repair work, cleaning, and **tuckpointing** old buildings. They can specialize in restoring historic brick buildings, which is a recognized craft specialty in some parts of the country. Masons have been around almost as long as farmers. Historically, they have been better paid.

Masonry is more than physical labor. It is a skilled occupation that calls for good hand-eye coordination, balance, and strength. It also requires good mental skills. This means ongoing study, concentration, and continued learning in an environment free from substance abuse.

Because masonry is highly skilled, it takes time to learn. Your learning starts with this course, combined with or followed by an apprentice's job. Masons work alone or as part of a team, so you will also learn to be a good team player. Masons work outdoors with a lot of lifting and bending, so you will learn to keep yourself in good shape. Masons work on high scaffolding, so you will learn safety rules and practices. Masons bring their skills and tools wherever they go.

5.1.0 CAREER STAGES

Masons were among the first workers to band together, or unionize. During the Middle Ages, the masons formed influential groups that still shape trade practices. Today, as in the past, masons' organizations recognize several stages of skill:

- Helper
- Apprentice
- Journeyman
- Foreman
- Superintendent
- Contractor

The helper is a common laborer, not a mason. The helper carries masonry, tools, and mortar and gets things for the mason. The helper mixes mortar, cleans tools, and gets to watch the mason at work. Sometimes, helpers decide they want to become masons. If they do, they must enter an apprenticeship program. Apprentices are at the beginning level of the masonry career path. Their training will lead them to full participation in the mason's trade and the opportunity for higher job levels.

The next sections discuss these different job levels.

5.2.0 APPRENTICE

An apprentice is a person who has signed an apprenticeship agreement with a local joint apprenticeship committee. The committee works with local contractors who have agreed to take apprentices. The U.S. Department of Labor regulates the apprenticeship process. In most states, the state department of labor is also involved, as state labor regulations provide guidelines on legal age requirements, pay, hours, and other aspects of an apprenticeship.

The length of the apprenticeship will vary. The U.S. Department of Labor program is three years and 4,500 hours with a minimum of 432 hours of classroom instruction. Manuals such as this NCCER series are used for classroom instruction or as part of apprenticeship training.

The apprentice is assigned to work with a contractor and to take classes. The apprentice must study and work under supervision. The apprentices agree to:

- Perform the work assigned by the contractor
- Abide by the rules and regulations of the contractor and the committee
- Complete the hours of instruction
- Keep records of work experience, training, and instruction
- Learn and use safe working habits
- Work with the assigned contractors for the entire apprenticeship period, unless reassigned by the committee
- Conduct themselves in an ethical manner, realizing that time, money, and effort are being spent to afford them this opportunity to become a skilled worker
- Remain free from drugs and alcohol abuse

A typical three-year apprenticeship is divided into six periods of six months each. The first six months is a trial period. The committee reviews the apprentice's performance and may end the agreement.

The apprentice attends classes and works under the supervision of a journeyer mason. As part of the supervised work, the apprentice learns to lay masonry units and perform other craftwork. The apprentice's pay increases for each six-month period as skill and performance increases.

At the end of the period, the apprentice receives a certificate of completion. This type of certificate, *Figure 27*, is known and accepted everywhere in the United States. The apprentice is now a journeyman mason.

5.3.0   JOURNEYMAN

Unlike an apprentice, a journeyman is a free agent who can work for any contractor. A journeyman can work without close supervision and is skilled in most tasks. The journeyman will find that the end of the apprenticeship is not the end of growing in the job.

The journeyman is a person with an excellent trade. Journeymen earn good wages in a trade that is always in demand. They have the satisfaction of creating and the opportunity to grow as masonry artists. They also have the opportunity to grow as layout persons, trainers, and supervisors.

An experienced and skilled journeyman can work as a layout person. For a pay premium, the layout person lays out the work and lays the leads. Less experienced masons and apprentices work between the leads set by the layout person. Experienced and skilled journeymen also train apprentices and supervise their work. With further experience, journeymen can supervise crews.

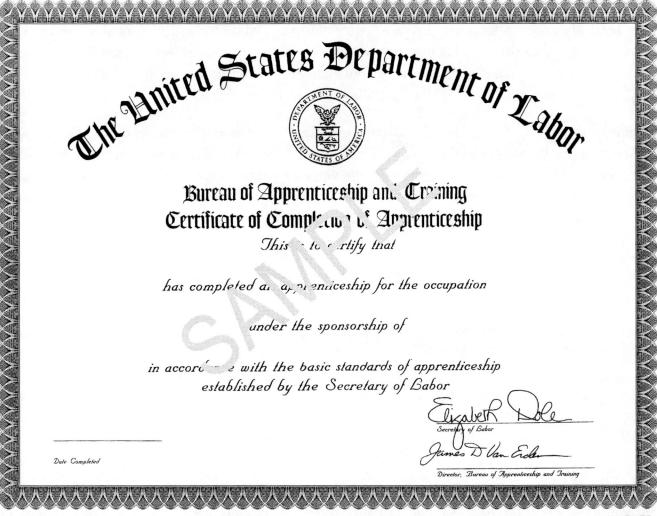

Figure 27. Example Apprenticeship Certification

Journeymen can continue to learn by studying and handling more complex tasks. They can continue to develop their skills as they work. Further education in masonry innovations and techniques is available and so is training in leadership and supervision.

5.4.0 SUPERVISORS, SUPERINTENDENTS, AND CONTRACTORS

Supervisors are responsible for managing and supervising a group of workers. This job requires a high degree of knowledge about masonry, as well as leadership and people-handling skills. Supervisors are typically responsible for training workers in safety measures and keeping work areas safe. They also train workers in new techniques or easier ways of working. They solve daily problems, keep on top of materials and supplies, and make sure workers meet job schedules. They check work to make sure it is done to standards. Supervisors may be called crew leaders, or forepersons, depending on the company that hires them.

Superintendents have several supervisors reporting to them. Usually, the superintendent is the lead person on a large job. For a smaller company, the superintendent may be in charge of all the work in the field for the contractor. The superintendent supervises the work of the supervisors and makes major decisions about the job under construction. The superintendent must have masonry skills, leadership skills, and business skills as well.

A masonry contractor may be like the master mason of the cathedral-building era: an artist, a sculptor, and a master builder, as well as a business owner. The contractors are responsible for the whole job, just as the master masons were centuries before. They bid on jobs, organize the work and the workers, inspect the work, confer with the clients, and run the business. The contractor needs to be able to plan and see ahead to keep up with change.

Contractors—and supervisors and superintendents—need to keep abreast of the latest knowledge and learn the newest ways of doing the work. Just as with apprentices, they need to keep on studying their trade.

6.0.0   KNOWLEDGE, SKILLS, AND ABILITY

Becoming a good mason takes more than being able to lay a masonry unit and level it. A competent mason is one who can be trusted to perform the required work and meet the project specifications. This mason must have the necessary knowledge, skills, and ability, as well as good attitudes about the work itself, about safety, and about quality.

6.1.0   KNOWLEDGE

Masons need to know how to handle all aspects of masonry work. They need to know how to:

- Read drawings and specifications, and interpret them
- Calculate and estimate quantities, lengths, weights, and volumes
- Select the proper materials for the job
- Lay masonry units into structural elements
- Work alone or as part of a team
- Assemble and disassemble scaffolding
- Keep tools and equipment in good repair and safe condition
- Follow safety precautions to protect themselves and other workers on the job

6.1.1   Job Site Knowledge

Masons need to be skilled in applying their knowledge to what they face each day on the job. The best way to do the work at a particular job site will depend on the layout of the work, what is happening around the masonry site, and the conditions surrounding the project.

Most of the mason's work is outside where the worker faces variations in temperature and weather. A mason must be able to work under these conditions and not be distracted by them. Masons must know how to react to changing conditions around them.

Much of this knowledge can be learned as you work, if you will pay attention. Notice what others do and ask questions. Ask your supervisor questions, too. Learn to respond to conditions at the job site. Masons must:

- Know the heart and the structure of their craft
- Have the skills to apply that knowledge to the work at hand
- Have the ability to do that work safely under all conditions

6.1.2 Learning More

Masons need to keep on learning after they finish their apprenticeships. They need to keep updating their skills all the time. The environment, tools, and expectations about masonry have evolved and will continue to change. Craftworkers and contractors alike will need to change the way they think about their work and how they do it.

National, regional, and local organizations offer continuing education for masons. Technical seminars, training sessions, publications, and classes are often free or low cost. They can bring you the latest information about tools, techniques, and new systems of working. To succeed, you must be alert to change and willing to learn new ways.

6.2.0 ATTITUDE AND WORK

Attitude can build an invisible bridge, or build an invisible wall, between us and others. No one wants to hang around a grouch or count on someone who is not dependable. No one minds helping someone who can do something in return or working with a friendly, cooperative partner. On top of knowledge, skills, and ability, the mason needs the right attitude. Your attitude comes from how you think and feel about your work and yourself.

6.2.1 Dependability

A mason must be dependable. Masonry work, like all construction, is a closely timed operation. Once started, it cannot stop without waste of material and money. Employers need workers who report to work on schedule. An undependable, absent worker will slow or stop masonry work and cost the project time and money. An undependable worker will not be able to depend on a job for very long.

6.2.2 Responsibility

Every masonry worker must be responsible for doing the assigned work in the proper manner. Every mason must be responsible for working safely by using skill, knowledge, and ability. The mason must be responsible enough to work without supervision and to work until the task is complete.

Being responsible for your own work includes admitting your mistakes. It also includes learning from your mistakes. Nobody is expected to be perfect. Everyone is expected to learn and to grow more skilled.

Employers are always in need of workers who are ambitious and want to become a leader or supervisor. Being responsible for what others do may be your career goal. The path to that goal starts with being responsible for what you do.

6.2.3 Adaptability

On any construction project, a large amount of work must be done in a short time. Planning and teamwork are needed to do so efficiently and safely. Supervisors sometimes form teams of two or more workers to do specific tasks. A mason may work in a team to erect a scaffold, then work alone for most of the day, then team with someone else to do a cleanup.

Masons on a job site can find themselves teaming with different people at different times. Being a team player becomes important. Team players accept instruction and direction. They communicate clearly. They keep an eye out for potential problems. They share information. They meet problems squarely with constructive ideas, not criticism.

All team players treat each other with respect. Everyone must be willing to work together. Everyone must be willing to bring their best attitude to the team. Team members need to be able to depend on each other. Team priorities ("Finish that task") must be more important than individual priorities ("I want to do something else"). People who are not willing to be team players may find themselves benched.

6.2.4 Pride

Pride in what you do comes from doing high-quality work in a timely manner and from knowing you are doing your best. Being proud of what you do can overflow into other areas. Proud workers take pride in their personal appearance. Their work clothes are clean, safe, neat, and suitable. Their clothes protect them from the elements and from falling masonry. Proud masons take pride in their tools. They have a complete set of well-maintained tools and other special equipment they need to do their jobs. They keep their tools safe and orderly in their masons' toolbags. They know how to use the right tool for the work at hand.

Proud masons work so that they can continue to be proud of what they do, and how they do it. Being proud of what you do is an important part of being proud of who you are.

6.3.0 SAFETY

Masons operate in a high-risk environment, in terms of safety. All around them are stacks of heavy materials, trucks, and loads going by on cranes. Work sites have many possibilities for accidents. Masons themselves can cause accidents. They can drop masonry or tools off scaffolding and onto other workers. They can assemble scaffolding so poorly that it collapses under the weight of the load. They can fall off scaffolding. They can use damaged or poorly maintained tools that injure themselves or others.

Masons must think and practice safety at all times. Their work must be planned so that it is safe as well as efficient.

All workers at a construction site must wear appropriate personal protective equipment. Masons need to protect their skin and eyes from mortar, grout, and flying masonry chips. They also need to protect themselves by being aware of what is happening around them. Masons need to keep track of the rest of their crew and of other crews. They need to keep part of their awareness on the lookout. Unusual movements or noises can indicate something is moving that should not be. Masons need to have the knowledge, skill, and ability to:

- Recognize an unsafe situation
- Alert fellow crew members to the danger
- Take evasive or corrective action

6.4.0 QUALITY

The latest ideas about quality in work are not new to masons at all. Those who work in masonry construction and finishing have been concerned about quality for over 10,000 years. The walls unearthed at Jericho were laid true and still stand true.

The quality of masonry depends on many factors. The mason building a wall may have little control over its design or the choice of masonry units. But the mason does have control over the quality of the completed job. A wall out of level, or with poorly finished joints, or plugged weepholes, is the mason's responsibility. Nothing stands between the mason's handiwork and the finished product except the mason's skill and commitment to quality.

The quality of the finished masonry structure depends directly on the knowledge, skill, and ability of the mason. Good work will be easily seen by all. Poor work will be seen even more easily. Given the durability of masonry, the quality of the work will be a monument to the skill of the mason for a very long time. The skilled, proud mason always strives for the highest quality that can be achieved.

SUMMARY

This module has introduced the history of masonry, which is traceable to thousands of years ago. Over the centuries, hand-formed dried mud brick has been replaced by molded mud brick, then by fired molded clay brick. Today, fired clay brick is available in a dazzling variety of shapes, sizes, colors, and textures.

Today's clay products are categorized as solid and hollow brick, structural and nonstructural tile, and made-to-order architectural terra cotta. Clay has been joined by cement as a modern masonry material. Cement masonry units outnumber clay units in their variety. Cement products are block, cement brick, and special-purpose block. Stone, the oldest recovered building material, is still laid by masons. Mortars and grouts have evolved from mud to special-purpose, high-strength cements.

Science and engineering have brought masonry into the modern age. With modern construction techniques, masonry is now used in high-rise buildings as well as residential and commercial projects. This allows for widespread use of masonry construction throughout North America.

Masonry as a career dates back to the ancient kingdoms of the Middle East. Masons were among the first craftworkers to organize into guilds. The guilds protected the secrets of the craft and maintained standards for the work. The legacy of the guilds includes the apprenticeship program available today.

Masonry as a career offers advancement through the recognized career steps of apprentice, mason, layout person, supervisor, superintendent, and masonry contractor. Success as a mason requires the willingness to keep on learning.

References

For advanced study of topics covered in this task module, the following reference materials are suggested:

Building Block Walls—A Basic Guide, National Concrete Masonry Association, Herndon, VA, 1988.

Masonry Design and Detailing—For Architects, Engineers, and Contractors, Fourth Edition, Christine Beall, McGraw-Hill Publishing, New York, NY, 1997.

Masonry Skills, Third Edition, Richard T. Kreh, Delmar Publishers, Inc., Albany, NY, 1990.

The ABCs of Concrete Masonry Construction, Videotape 13:34 min, Portland Cement Association, Skokie, IL, 1980.

REVIEW/PRACTICE QUESTIONS

1. Mortar, dry stacking, and mechanical connectors _____.

   a. are historic masonry techniques
   b. connect adobe to ashlar or rubble
   c. can substitute for grout
   d. are steps in brickmaking

2. Early wooden molds for shaping clay brick led to the _____.

   a. development of pattern walls
   b. standardization of brick sizes
   c. use of mortar
   d. use of kilns

3. The ligier was the medieval equivalent of a(n) _____.

   a. trowel
   b. composite wall
   c. apprentice
   d. running bond

4. The first fired brick buildings in North America were built during _____.

   a. the 1930s
   b. the early 1900s
   c. colonial times
   d. the Civil War

5. In the 1940s, engineers developed formulas for _____.

   a. masonry bearing walls
   b. fireproof concrete block
   c. reinforcing masonry
   d. new mortars

6. Structural clay products include _____.

   a. brick, block, and tile
   b. brick, tile, and terra cotta
   c. tile, terra cotta, and prefaced units
   d. block, grout, and tile

7. During the 1960s, concrete block was _____.
   a. brought under ASTM standards
   b. made of cinders and fly ash
   c. used in more walls than clay brick
   d. cured by high-pressure steam

8. Rubble stone is _____.
   a. slump brick
   b. made of concrete
   c. cut with smooth bedding surfaces
   d. irregular in size and shape

9. Modern mortar is made of _____.
   a. cement, lime, and sand
   b. cement, sand, admixtures, and water
   c. sand, lime, ceramic, and water
   d. cement, sand, and water

10. Solid masonry units have voids occupying _____.
    a. less than 75 percent of their surface
    b. less than 25 percent of their surface
    c. more than 25 percent of their mass
    d. more than 50 percent of their mass

11. Veneer walls are _____.
    a. built 6½ inches away from the weight-bearing wall
    b. commonly used in industrial construction
    c. tied to the loadbearing wall with metal ties
    d. not designed for appearance

12. The single thing most affecting how masons work today, as compared to ancient times, is the _____.
    a. lack of a guild
    b. uniformity of ASTM standards and specifications
    c. shortened time to become a master mason
    d. use of power tools and equipment to make the work easier

13. With all the advancements of the last 100 years, masonry _____.
    a. is still very much a craft and an art
    b. can be practiced by anyone
    c. is not important in construction work
    d. still requires 10 years of training

14. Masons can find work today on projects for building everything but _____.

    a. large commercial or high-rise buildings
    b. home building, repair, or additions
    c. historic renovation and reconstruction
    d. cathedrals and fortresses

15. Masonry career stages today include _____.

    a. foreman
    b. ligier
    c. master
    d. restorer

16. Apprentice masons agree to _____.

    a. keep records of what they have learned
    b. stay in the apprenticeship until it is complete
    c. attend classes instead of working
    d. cause accidents

17. Today's mason with the skills closest to the master mason would be the _____.

    a. apprentice
    b. masonry contractor
    c. layout artist
    d. supervisor

18. Today's masons need to know how to _____.

    a. draw up construction plans
    b. calculate and estimate quantities, weights, and volumes
    c. operate lift equipment
    d. omit safety precautions

19. Even though they may work alone, masons do need _____.

    a. knowledge of safety procedures
    b. to show up on time
    c. to know what is happening around them
    d. good teamwork skills

20. Masons can avoid accidents by _____.

    a. taking shortcuts
    b. paying attention to what is happening around them
    c. assembling scaffolding quickly
    d. wearing loose clothing and tennis shoes

ANSWERS TO REVIEW/PRACTICE QUESTIONS

<u>Answer</u>		<u>Section Reference</u>
1.	a	1.0.0
2.	b	2.1.0
3.	c	2.2.0
4.	c	2.3.0
5.	a	2.3.0
6.	b	3.1.0
7.	c	3.2.0
8.	d	3.3.0
9.	b	3.4.0
10.	b	4.1.0
11.	c	4.1.0
12.	d	4.2.0
13.	a	4.2.0
14.	d	5.0.0
15.	a	5.1.0
16.	b	5.2.0
17.	b	5.4.0
18.	b	6.1.0
19.	d	6.2.3
20.	b	6.3.0

ACKNOWLEDGMENTS

Figures 1 and 24 courtesy of McGraw-Hill Publishing

Figures 2, 8, and 22 courtesy of Roy Jorgensen Associates, Inc.

Figures 3 and 16 courtesy of National Concrete Masonry Association

Figures 4-6, 12, and 14 courtesy of U.S. Brick Company

Figure 7 courtesy of Earthlore

Figure 9 courtesy of Lighthouse Society

Figure 10 courtesy of Chicago Historical Society

Table 1 courtesy of Carolina Concrete Masonry Association

Figures 11, 17, and 23 courtesy of The Goodheart-Willcox Company Inc.

Figures 13 and 15 courtesy of Dover Books

Figures 18-21 and Table 2 courtesy of Portland Cement Association

Figures 25 and 26 courtesy of Associated General Contractors

Figure 27 courtesy of U.S. Department of Labor

The NCCER makes every effort to keep these manuals up-to-date and free of technical errors. We appreciate your help in this process. If you have an idea for improving this manual, or if you find an error, a typographical mistake, or an inaccuracy in the NCCER's Craft Training Manuals, please write us, using this form or a photocopy. Be sure to include the exact module number, page number, a description of the problem, and the correction, if possible. Your input will be brought to the attention of the Technical Review Committee. Thank you for your assistance.

Instructors – If you found that additional materials were necessary in order to teach this module effectively, please let us know so that we may include them in the Equipment/Materials list in the Instructor's Guide.

Write: Curriculum and Revision Department
National Center for Construction Education and Research
P.O. Box 141104
Gainesville, FL 32614-1104
Fax: 352-334-0932

Craft _____ Module Name _____

Module Number _____ Page Number(s) _____

Description of Problem

_____

_____

_____

_____

(Optional) Correction of Problem

_____

_____

_____

(Optional) Your Name and Address

_____

_____

_____

notes

Safety Requirements

Module 28102

SAFETY REQUIREMENTS

NATIONAL
CENTER FOR
CONSTRUCTION
EDUCATION AND
RESEARCH

OBJECTIVES

Upon completion of this module, the trainee will be able to:

1. Describe safety precautions and general housekeeping practices that should be followed at a typical work site.

2. Describe the safety precautions that should be followed when working in special areas such as trenches, excavations, confined spaces, scaffolding, and limited access zones.

3. Describe the proper procedures for handling and maintaining masonry tools safely.

4. Explain the importance of safety meetings and what they involve.

5. Identify and discuss the purpose of federal safety designation colors.

6. Demonstrate setting up ladders according to OSHA safety regulations.

7. Discuss the uses of and demonstrate proper procedures for putting on eye protection, respiratory protection, and a safety harness.

8. Demonstrate correct safety procedures for fueling and starting a gas-fueled power tool.

Prerequisites

Successful completion of the following Task Modules is recommended before beginning study of this Task Module: Core Curricula; Masonry Level 1, Module 28101.

Required Trainee Materials

1. Trainee Task Module
2. Pencil and paper
3. Appropriate Personal Protective Equipment

COURSE MAP

This course map shows all of the task modules in the first level of the Masonry curricula. The suggested training order begins at the bottom and proceeds up. Skill levels increase as a trainee advances on the course map. The training order may be adjusted by the local Training Program Sponsor.

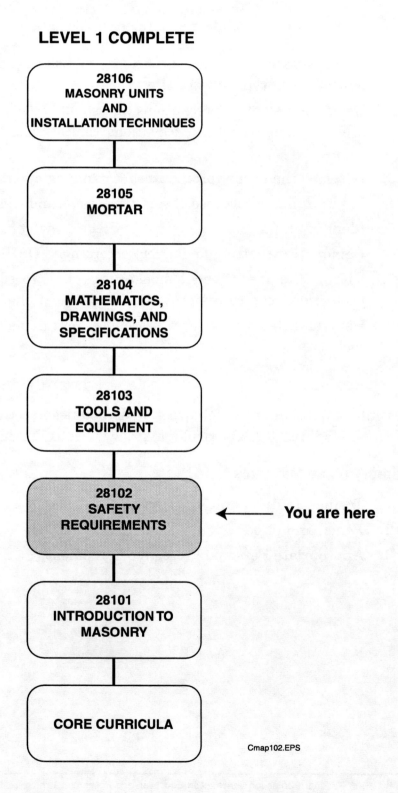

LEVEL 1 COMPLETE

28106
MASONRY UNITS
AND
INSTALLATION TECHNIQUES

28105
MORTAR

28104
MATHEMATICS,
DRAWINGS, AND
SPECIFICATIONS

28103
TOOLS AND
EQUIPMENT

28102
SAFETY
REQUIREMENTS   ← **You are here**

28101
INTRODUCTION TO
MASONRY

CORE CURRICULA

Cmap102.EPS

TABLE OF CONTENTS

Trade Terms Introduced In This Module

Accident: An unexpected, unintentional event that causes injury, death, or property damage.

Alkaline: Bitter, slippery, or caustic.

ANSI: American National Standards Institute; a private organization that develops and promotes standards to be used throughout industry and commerce.

Carelessness: Inattentiveness; failure to pay attention.

Caustic: The ability to burn, corrode, dissolve, or eat away.

Confined space: Any area with limited exits that has toxic or flammable materials or limited amounts of air to breathe, such as storage tanks, sewers, and tunnels.

Extension ladder: Two straight ladders assembled such that the overall length of the combination can be adjusted.

Hazard: A source of danger or potential accident.

Hygroscopic: A tendency to absorb moisture readily from the skin.

Lanyard: A short section of rope used to attach a worker's safety harness to a strong anchor point located above the work area.

OSHA: The Occupational Safety and Health Administration; an agency of the U.S. Department of Labor.

Personal Protective Equipment: Pieces of equipment or clothing designed to prevent or reduce the effect of injuries.

Respirator: A device designed to provide clean, filtered air for breathing no matter what is in the surrounding air.

Safety: Freedom from danger, risk, or injury.

Scaffold: An elevated work platform for both personnel and materials.

Scaffolding: A manufactured or job-built structure that supports a work platform.

Stepladder: A self-supporting ladder consisting of two elements that are hinged at the top.

Toxic: Harmful, destructive, deadly, or poisonous.

1.0.0 INTRODUCTION TO SAFETY

As introduced in the Core Curricula module, *Basic Safety*, **safety** is an attitude. It is a way of working on the job. The time spent learning and practicing safety procedures can literally save your life—and the lives of others. According to the National Safety Council, the organized safety movement has saved more than three million lives so far, thereby preventing three million tragic deaths.

The final responsibility for on-the-job safety rests with each and every worker. It is every worker's responsibility to learn:

- the safety gear and clothing required for each job
- the safety rules and procedures associated with each job
- general work site safety and the **hazards** associated with each job
- materials handling safety as it affects each job

This module will present this information, as well as safety tips for working with and maintaining ladders, **scaffolding**, and tools. The procedures, practices, and rules presented in this module will help you work safely so that you can identify and avoid hazardous conditions on the job site. By practicing safe procedures and keeping a safe attitude, you will do your part to keep your workplace free from **accidents** and protect yourself and your fellow workers from harm.

1.1.0 PERSONAL RESPONSIBILITY

Skilled masonry craftworkers must be constantly aware of threats to their safety and to the safety of others in the work area. The masonry profession requires workers to use many dangerous tools and work in areas where an accident is possible with even momentary **carelessness**.

Studies show that there is a much greater risk for injury when workers become comfortable in performing skills because they tend to become more careless. Practicing safe procedures and developing safe habits can make safety your natural reaction, which can help prevent injuries, loss of equipment, and loss of materials. It may even save lives.

On-the-job safety is a concern not only because many companies require safe work records, but, more importantly, because safe working practices and conditions ensure a safe operation. This, in turn, helps to ensure the safety and well-being of the workers. Thinking before you act, staying alert, and being aware of where you are and what you are doing can help prevent personal injury. It can also help to ensure that you will be able to continue working to provide for your family and keep them safe.

Working responsibly and safely can also help prevent accidents and harm to your fellow workers. This is why it is so important to encourage everyone involved in a job or working on a job site to be safety conscious. The use of safe working practices also makes your employer happy because it means that all the workers will be able to work and that the job will progress smoothly and finish on time and within budget.

1.2.0 COST OF JOB ACCIDENTS

Unsafe working conditions and practices can result in:

- Personal injury or death
- Injury or death of other workers
- Damage to equipment
- Damage to the work site

According to the Occupational Safety and Health Administration (**OSHA**), there were more than 6,000 work-related deaths in 1996. Aside from the pain and suffering for the individuals involved and their families, injury and death have ramifications for others as well. For example, accidents can slow down or stop a job, thereby putting the entire operation in jeopardy, and possibly resulting in site-wide layoffs. A high accident rate can cause an employer's insurance rates to increase, making the company less competitive with other construction companies and less likely to secure future work.

Damage to the equipment and work site also affects more than just the one job. Damage to the equipment can either stop a job for a period of time until the equipment can be fixed, or increase the cost of a job because other machinery has to be brought in. Damage to the work site will also increase the cost of the job, requiring repair and rework. This is why safety must be made a part of every job and a part of everyone's responsibility.

Each year, construction companies lose thousands of dollars repairing or replacing equipment and dealing with production slowdowns and other hidden costs that are caused by accidents. And when your employer loses money, so do you. The money that would have been available for raises and bonuses is easily taken away by the cost of a serious on-the-job accident.

On the other hand, a safe attitude and an effective safety program can help keep everyone in business. To ensure a safe work site, all workers must familiarize themselves with and follow all OSHA safety standards, general industry safety orders, and company-specific safety rules. Safety posters and other safety materials are displayed for your benefit. Make sure you read them and follow their suggestions.

1.3.0 CAUSE OF ACCIDENTS

In today's environment, there are several causes of accidents. These include the workers themselves, the tools and equipment used on the job, and the job site. In order to work safely and securely, the workers must be in good physical and mental condition. They should be free from stress, drugs, and alcohol, and be focused on the job at hand. The tools and equipment used on the job must be well-maintained and in good working condition, and the operator must know how to use them appropriately. The job site should also be clean and free of any hazardous materials or conditions.

Alcohol or drug use on the job cannot be tolerated because drug or alcohol users present a safety threat not only to themselves but also to their fellow workers. The use of alcohol or drugs on the job causes impaired perception as well as impaired mental and physical abilities. It also results in tardy or absent workers who neglect their duties and are less productive. Alcohol and drug use on the job also drives up the cost of production and the cost of insurance, making employers less competitive.

Additional causes of accidents include a failure to communicate clearly, poor work habits, lack of Appropriate **Personal Protective Equipment**, and improper work attire. It is important to communicate your intended actions so that those working with or near you understand what you will be doing. It is also important for you to question what others are doing if you are unsure or uncertain. Using good communication and good work habits will help you create a safe environment to work in. Wearing the appropriate protective gear and clothing for the job you are performing will also help keep you safe. More detailed information regarding safe working practices and conditions is presented throughout this module.

2.0.0 GENERAL JOB SITE SAFETY

The Occupational Safety and Health Administration (OSHA) makes rules and regulations regarding the physical environment of the workplace. Yet, physical hazards on the job site have resulted in more accidents and damages than any other aspect of working in construction. According to OSHA, falls are the leading cause of worker fatalities in the construction industry in the United States. Each year, on average, between 150 and 200 workers are killed and more than 100,000 are injured as a result of falls at construction sites.

Problems at the job site are both environmental and human. Accidents may be caused by unsafe equipment operation, unsafe working conditions, or unsafe clothing. If the conditions are a little unsafe, and if the workers do not work extra carefully, the whole situation becomes doubly unsafe. For example, open construction sites are naturally hazardous. As a building frame goes up, things are likely to fall. Another example of a hazardous situation has to do with higher working heights. When you are working on the fourth or fifth floor, for instance, there are open holes you can fall through. You must be extra cautious. You must also think.

If something starts to fall over the side of the building, let it go. Shout "heads up" and stay put. Do not be a hero and try to balance on the edge of the fifth floor to keep something from going over. No matter how expensive, tools or materials can be replaced. You, however, will not be able to recover quite so easily from a fall.

2.1.0 SAFETY OBLIGATIONS

In order to be safe, you need to work safe and work smart. This includes conducting yourself in a manner that will ensure safe practices; being conscious of others' safety and encouraging others to do the same; and familiarizing yourself with all federal, state, and local laws, company policies and procedures, manufacturer recommendations, and job-specific safety rules. By your actions, you can prevent accidents and injuries and take pride in your accomplishments on the job, in the industry, and at home.

You need to make it your business to take a serious look at safety. Construction work is dangerous. Not only is it easy to fall off a roof or cut a finger, but it is likely that you *will* unless you take safety precautions. Think about this: all it takes is one serious accident—an accident that can happen in a fraction of a second—and you can ruin the rest of your life. You owe it to yourself, your family, your employer, and your fellow workers to learn all you can about safety and to work safely.

Safety procedures must be learned for each task a worker does. Sometimes, new material or equipment is brought on the job and new safety procedures must be learned by everyone on the job. The important thing to remember is that safety *can* be learned. What you learn about safety can help prevent accidents.

It is also important to understand the obligations that exist for everyone's safety. An obligation is like a promise or a contract. In exchange for the benefits of your employment, you agree to work safely. In other words, you are obligated to work safely. You are also obligated to make sure that anyone you supervise is working safely. Finally, of course, your employer is obligated to maintain a safe workplace for all employees.

Some employers will have safety committees. If you work for such an employer, you are then obligated to that committee to help maintain a safe working environment. This means two things:

- Follow the safety committee's rules for proper working procedures and practices.
- Report any unsafe equipment and conditions directly to the committee or a committee member, or to your supervisor.

If you see something that is not safe, report it. Do not ignore it. It will not become safe again all by itself. You have an obligation to report it. In the long run, even if you do not think an unsafe condition affects you—it does. So, report what is not safe. Do not think that your employer will be angry because your productivity slows down while you report the condition. On the contrary, your employer will be very pleased.

Your employer knows very well that the short time lost in making it safe again is nothing compared with shutting down the whole job because of a major disaster. And if that happens, you are out of work anyway. So, do not ignore an unsafe condition. In fact, federal regulations under OSHA require you to report unsafe and hazardous conditions. If you are worried about your job being on the line, think about it in terms of your life being on the line.

2.2.0   OSHA STANDARDS

In 1970, the U.S. Congress passed the Occupational Safety and Health Act in order to ensure safe and healthful working conditions for the more than 90 million working men and women in the nation. This national safety and health law establishes standards that require employers to provide their workers with workplaces that are free from recognized hazards that could cause serious injury or death. It also requires employees to comply with all safety and health standards that apply to their jobs. One of the major outcomes of this Act was the establishment of the OSHA in the Department of Labor, which enforces these standards.

According to OSHA standards, you are entitled to on-the-job instruction. Specifically, every new employee must be:

- Shown how to do their job safely
- Provided with the Appropriate Personal Protective Equipment
- Warned about specific hazards in the work and in the surroundings
- Supervised for safety while performing the work

The enforcement of this Act is provided by federal and state safety inspectors who have the legal authority to make employers pay fines for safety violations. The law provides that states may have their own safety regulations and agencies to enforce them, but they must first be approved by the U.S. Secretary of Labor. For states that do not develop such regulations and agencies, the law says that federal OSHA standards must be obeyed.

These standards have been published in a book known as the *OSHA Safety and Health Standards* (29 CFR 1926), which is sometimes called simply "OSHA Standards 1926."

The most important general requirements that OSHA makes on employers in the construction industry are these:

- The employer must initiate and maintain whatever programs are necessary to provide for frequent and regular job site inspections of equipment.
- The employer must instruct all employees to recognize and avoid unsafe conditions and to know the regulations that pertain to the employees so they may control or eliminate any hazards of injury.
- No one may use any tools, equipment, machines, or materials that do not comply with OSHA Standards 1926.

2.2.1 Scaffolding

OSHA has very strict requirements regarding the equipment that should be used around a construction site. Section 1926.451, for example, regulates the use of scaffolding, including what type of wood should be used for planking, what size the planks should be, how much space the planks should span, and how far the planks can extend over their end supports. OSHA also regulates the type of rope that can be used to suspend **scaffolds**.

In addition to general material requirements and definitions, OSHA has specific requirements for working with and using different types of scaffolding, including wood pole scaffolds, tube and coupler scaffolds, tubular welded frame scaffolds, manually propelled mobile scaffolds, elevating and rotating work platforms, outrigger scaffolds, masons' adjustable multiple-point suspension scaffolds, swinging scaffolds, plank-type platforms, ladder-type platforms, beam-type platforms, and light metal-type platforms.

General safety guidelines for working with scaffolding include the following items:

- Make sure the footing or anchorage for scaffolds is sound, rigid, and capable of carrying the maximum intended load without settling or displacement. Do not use unstable objects such as barrels, boxes, loose brick, or concrete blocks to support scaffolds or planks.

- Do not erect, move, dismantle, or alter scaffolds except under the supervision of competent persons.

- Where persons are required to work or pass under the scaffold, provide the scaffold with a screen between the toeboard and the guardrail, extending along the entire opening, consisting of No. 18 gauge U.S. Standard wire ½-inch mesh, or the equivalent.

- Make sure the scaffold and its components are capable of supporting at least four times the maximum intended load.

- Repair or replace any scaffold or accessories such as braces, brackets, trusses, screw legs, ladders, etc., that are damaged or weakened.

- Provide an access ladder or equivalent safe access.

- Provide overhead protection for workers on a scaffold that is exposed to overhead hazards.

- Eliminate all slippery conditions on scaffolds as soon as possible after they occur.

- Do not perform any welding, burning, riveting, or open flame work on any staging that is suspended by means of fiber or synthetic rope.

- Do not use shore or lean-to scaffolds.

- Make sure all materials such as palletized masonry units being hoisted onto a scaffold have a tag line.

- Do not work on scaffolds during storms or high winds.

- Do not allow tools, materials, or debris to accumulate in quantities that can cause a hazard.

OSHA also has training provisions regarding scaffolding. For example, OSHA mandates that the employer should provided a training program for each employee using scaffolding. The program should enable each employee to recognize hazards related to scaffolds and should train each employee in the procedures to be followed to minimize these hazards. The employer should ensure that each and every employee has been trained by a competent person in the following areas:

- The nature of fall hazards

- The correct procedures for erecting, maintaining, and disassembling the fall protection system to be used

- The proper assembly, maintenance, inspection, and disassembly of all scaffolding used on the job site

2.2.2 Ladders

Sections 1926.1051 and 1053 include general safety guidelines for working with ladders. These sections tell when a ladder must be used at a construction project and what type or types of ladder should be used. For example, when ladders are the only means of access or exit from a working area for 25 or more employees, or when a ladder is to serve simultaneous two-way traffic, a double-cleated ladder or two or more separate ladders must be provided. When a building or structure has only one point of access between levels, that point of access must be kept clear to permit free passage of employees. When work must be performed or equipment must be used such that free passage at that point of access is restricted, a second point of access must be provided and used. When a building or structure has two or more points of access between levels, at least one point of access must be kept clear in order to permit free passage of employees.

Additional guidelines discuss the load capacity of different types of ladders, as well as the construction of ladders. Ladder rungs, cleats, and steps, for example, should be parallel, level, and uniformly spaced when the ladder is in position for use. Rungs, cleats, and steps of portable ladders and fixed ladders should be spaced not less than 10 inches (25 cm) apart, nor more than 14 inches (36 cm) apart, as measured between center lines of the rungs, cleats, and steps. Rungs, cleats, and steps of step stools should be spaced not less than 8 inches (20 cm) apart, nor more than 12 inches (31 cm) apart, as measured between center lines of the rungs, cleats, and steps. Ladder components must be surfaced so as to prevent injury to an employee from punctures or lacerations and to prevent snagging of clothing. The rungs and steps of ladders should also be surfaced to minimize slipping.

Non-self-supporting ladders should be used at an angle such that the distance from the top support to the foot of the ladder is approximately one quarter of the working length of the ladder. The working length is defined as the distance along the ladder between the foot and the top support. Wood job-made ladders with spliced side rails should be used at an angle such that the horizontal distance is one eighth the working length of the ladder. Fixed ladders should be used at a pitch no greater than 90 degrees from the horizontal, as measured to the back side of the ladder.

Ladders should be used only on stable and level surfaces unless secured to prevent accidental displacement. The top of a non-self-supporting ladder should be placed with the two rails supported equally unless it is equipped with a single support attachment. When portable ladders are used for access to an upper landing, the ladder side rails should extend at least 3 feet above the upper landing surface. When such an extension is not possible because of the ladder's length, then the ladder should be secured at its top to a rigid support, and a grasping device, such as a grabrail, should be provided.

Ladders should not be moved, shifted, or extended while occupied. When ascending or descending a ladder, the mason should face the ladder and use at least one hand to grasp the ladder. When climbing ladders, workers should not carry any object or load that could cause them to lose balance and fall.

OSHA also has training provisions regarding ladders. For example, OSHA mandates that the employer should provide a training program for each employee using ladders. The program should enable each employee to recognize hazards related to ladders and should train each employee in the procedures to be followed to minimize these hazards:

- The nature of fall hazards in the work area
- The correct procedures for erecting, maintaining, and disassembling the fall protection system to be used
- The proper construction, use, placement, and care in handling of all stairways and ladders
- The maximum intended load-carrying capacities of ladders used

Retraining should be provided for each employee as necessary so that employees maintain their understanding and knowledge.

2.2.3 Trenches And Excavations

Sections 1926.650 to 1926.652 cover safety guidelines and requirements for trenches and excavations. The OSHA rules regarding trenches and excavations apply to all open excavations made in the earth's surface. OSHA defines an excavation as "any man-made cut, cavity, trench, or depression in an earth surface, formed by earth removal." It defines a trench as "a narrow excavation (in relation to its length) made below the surface of the ground. In general, the depth is greater than the width, but the width of a trench (measured at the bottom) is not greater than 15 feet."

OSHA's specific excavation requirements include information regarding surface encumbrances, underground installations, access and egress, exposure to falling loads, warning systems for mobile equipment, hazardous atmospheres, and protection from hazards associated with water accumulation.

Under access and egress, for example, this section presents specifications regarding structural ramps, stating that if the ramps and runways are constructed of two or more structural members, they must be connected together so as to prevent displacement. The structural members used for ramps and runways should also be of uniform thickness. Under means of egress from trench excavations, this section presents information regarding stairways, ladders, ramps, or other safe means of egress. According to OSHA requirements, means of egress should be located in trench excavations that are 4 feet (1.22 m) or more in depth so as to require no more than 25 feet (7.62 m) of lateral travel for employees.

Section 1926.652 provides specific requirements for protective systems that must be used when working in trenches and excavations. Each employee in an excavation should be protected from cave-ins by an adequate protective system, except when excavations are made entirely in stable rock or when excavations are less than 5 feet (1.52m) in depth and examination of the ground by a competent person provides no indication of a potential cave-in. Protective systems should have the capacity to resist without failure all loads that are intended or could reasonably be expected to be applied or transmitted to the system.

This section also covers different designs of sloping and benching systems, support systems, shield systems, and other protective systems; materials and equipment used for protective systems; and the installation and removal of support, sloping and benching, and shield systems.

2.2.4 Appropriate Personal Protective Equipment

OSHA also has very strict standards regarding personal protective equipment. In general, the employer is responsible to OSHA in making sure that all employees wear the Appropriate Personal Protective Equipment whenever those employees are exposed to possible hazards to their safety. This includes strict guidelines regarding the use of approved ear protection devices when exposed to high sound levels and the use of approved **lanyards** and lifelines when working in dangerous situations above the ground.

Sections 1926.95 to 1926.106 state that protective equipment, including Appropriate Personal Protective Equipment for eyes, face, head, and extremities; protective clothing; respiratory devices; and protective shields and barriers, shall be provided, used, and maintained in a sanitary and reliable condition in order to protect workers from injury or impairment caused by environmental or process hazards, chemical hazards, radiological hazards, or mechanical irritants.

The specific type of equipment that should be worn is described later in this module.

2.2.5 Hazardous Materials

OSHA defines a hazardous substance as "a substance which, by reason of being explosive, flammable, poisonous, corrosive, oxidizing, irritating, or otherwise harmful, is likely to cause death or injury." Section 1926.59 provides a very thorough description of OSHA's hazard communication requirements, including ways to respond to emergency situations. A written plan for emergency situations, for example, should be developed for each construction operation where there is a possibility of an emergency. The plan should include procedures where the employer identifies emergency escape routes for the employees at each construction site before the construction operation begins.

Procedures for monitoring exposure to hazardous materials are also provided, as is the procedure for establishing regulated areas where airborne concentrations of hazardous materials exceed, or can reasonably be expected to exceed, the permissible exposure limits. Concentrations of mineral dust from cutting brick or block may become hazardous over time and lead to silicosis. Appropriate Personal Protective Equipment is required to prevent this hazard.

2.2.6 Federal Safety Designation Colors

With so many various hazards in the construction industry, the government developed a color system to readily identify hazards and dangers in the workplace. Employers are required to use this color system to mark dangerous or hazardous equipment or areas in their workplace. A chart depicting the system is shown in *Figure 1*.

You should study this chart and familiarize yourself with the various colors and what they mean. If you are colorblind, inform your supervisors so they can determine some other way to teach you this information.

2.2.7 Fall Protection

OSHA has revised its construction industry safety standards and developed systems and procedures that are designed to prevent employees from falling off, onto, or through working levels and to protect employees from being struck by falling objects. The performance-oriented requirements make it easier for employers to provide the necessary protection.

Section 1926.501 identifies areas or activities where fall protection is needed. These include, but are not limited to, ramps, runways, and other walkways; excavations; hoist areas; holes; formwork and reinforcing steel; leading edge work; unprotected sides and edges; overhand bricklaying and related work; roofing work; precast concrete erection; wall openings; residential construction; and other walking/working surfaces. The rule sets a uniform

COLOR	DESIGNATION	IDENTIFICATION
Federal safety red	Danger	Fire-protection equipment and its location Portable containers of flammable liquids Emergency stop bars, stop buttons, and emergency electrical stop switches on machinery
Federal safety yellow	Caution, particulary against starting, using, or moving equipment while it is under repair, and against the use of defective equipment	Waste containers for explosive or combustible equipment Starting point or power source of machinery Physical hazards such as low beams, protruding extensions, and steps
Federal safety orange	Dangerous parts of equipment that may cut, crush, shock, or otherwise injure	Safety starter buttons and parts of equipment that may cause electrical shock Exposed parts (edges only) of pulleys, gears, rollers, cutting devices, and power jaws
Federal safety purple (or black or magenta on safety yellow)	Radiation hazards	
Federal safety blue	Caution; provides information about machines or equipment that is out of order or under repair	Out-of-order signs on equipment
Federal safety green	The location of safety equipment other than fire-fighting equipment; the location of first-aid equipment	
Federal safety black and white (used individually or in combination)	Traffic flow	Items for housekeeping purposes Storage areas

102F01.EPS

Figure 1.   Color Designation And Identification System

threshold height of 6 feet, thereby providing consistent protection. This means that construction employers must protect their employees from fall hazards and falling objects whenever an affected employee is 6 feet or more above a lower level. Protection must also be provided for construction workers who are exposed to the hazards of falling into dangerous equipment and working around falling objects.

Fall protection generally can be provided through the use of guardrail systems, safety net systems, personal fall arrest systems, positioning device systems, and warning line systems, among others.

Full body harnesses with lanyards should be used for restraint and/or fall arrest where vertical free-fall hazards exist, and to reduce the probability of falls. Lanyards should limit the fall to 3 feet. If the lanyard allows the worker to drop more than 3 feet, a shock-absorbing system should be employed. Where a possibility exists for a vertical free-fall of more than 3 feet, body harnesses and lanyards should be used for suspension, support, and or positioning, and a shock-absorbing system should be used.

Section 503 provides training requirements that state that the employer shall provide a training program for each employee who might be exposed to fall hazards. The program should enable each employee to recognize the hazards of falling and should train each employee in the procedures to be followed in order to minimize these hazards.

2.3.0 SAFETY MEETINGS

Safety begins with proper training and orientation. Newly hired operators should not be allowed to operate any equipment until they have read and demonstrated an understanding of the company's safety program. In addition, workers who are being trained as operators should not perform work in unusually hazardous areas or operate equipment until they have received a safety orientation and checkout from a supervisor.

To reinforce safety policy and procedures as well as inform workers of any new safety standards, your employer should hold regularly scheduled safety meetings. Short topical meetings at the job site are usually held to keep attention focused on the safety concerns. These short tailgate or "tool box" meetings are often held first thing in the morning when everyone is alert and before people have a chance to spread out. Topics include accident prevention, safe working practices, emergency procedures, tool and equipment operation, and personal protective gear. The meeting discussions and demonstrations are led by a safety officer or supervisor who is qualified to present topical safety material or information. These meetings are usually documented on a form like the example shown in *Figure 2*; everyone attending the meeting signs the form.

In addition to tailgate meetings, companies often have formal training programs that operators are required to attend on a periodic basis. Topics of these meetings include first aid, emergency procedures, OSHA and other government safety requirements and regulations, hazardous materials handling, and licensing and certification requirements. It is your responsibility to attend these meetings and pay attention to what is being presented.

Many people do not like to go to meetings or training sessions because they think they are a waste of time. Others are always eager to attend because they think they are getting out of work and relaxing at the company's expense. However, meetings, especially training sessions, are work and should be treated like any other part of your job. Some meetings may be boring, but try to focus on the reason you are there and learn everything you can about the subject being discussed. In the case of safety information, it might save your life.

```
+-------------------------------------------------------+
|                                                       |
|              TAILGATE SAFETY MEETING                  |
|               General Safety Discussion               |
|                                                       |
|     Job No. _____   Date _____   Supervisor's Name _____  |
|                                                       |
|     SUPERVISOR'S COMMENTS OR INSTRUCTIONS             |
|                                                       |
|                                                       |
|                                                       |
|     CRAFTWORKERS' COMMENTS                            |
|                                                       |
|                                                       |
|                                                       |
|     EMPLOYEE NUMBERS AND INITIALS OF THOSE IN ATTENDANCE  |
|                                                       |
|                                                       |
|                                                       |
|     _____                         |
|     Supervisor's Signature                            |
|                                                       |
|              GF Review _____   Superintendent Review _____  |
|   When completed, return to your General Foreman. After review, return to Safety Department  |
+-------------------------------------------------------+
```

102F02.EPS

Figure 2. Tailgate Meeting Documentation

2.4.0 PERSONAL SAFETY AND PROTECTIVE EQUIPMENT

Part of safety training involves familiarizing yourself with the appropriate safety gear and clothing for the job. Safety gear and clothing, when worn appropriately, can help protect you from harm. The type of clothing and equipment required varies with job responsibilities and daily activities. It is the responsibility of each and every worker to see that they wear the appropriate clothing for the task they are performing. Personal safety equipment is designed to create a barrier against workplace hazards. It is your duty to use this equipment and use it properly. Your supervisor will provide a specific list of safety clothing and equipment required for your job.

2.4.1 Types Of Protective Equipment

As mentioned earlier, OSHA has very strict guidelines regarding the types of protective equipment that should be used. In addition, safety devices must be approved by the American National Standards Institute (**ANSI**), a recognized national technical association that publishes standards and specifications for materials and devices. The general types of safety clothing and equipment used by masons are described in the following paragraphs.

A safety helmet or hard hat, as shown in *Figure 3*, is used to protect the worker's head from any falling or flying materials or from bumps to the head. Construction sites are filled with activity from all directions. Therefore, it is important to wear a safety helmet or hard hat at all times, especially when working under ladders or scaffolding.

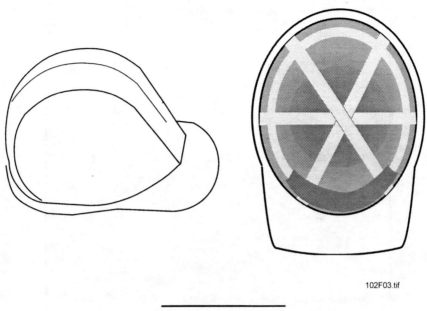

102F03.tif

Figure 3. Hard Hat

Hard hat components consist of a shell, suspension, and headband. The best hard hats are made of rigid plastic and have an inner lacing of webbing. This can be adjusted, but it should not be removed because it is what enables the hard hat to absorb shock. Ventilation is provided by a space between the suspension headband and the shell.

All hard hats must meet the specifications contained in ANSI Z89.1, *Safety Requirement for Industrial Head Protection*. No structural or design changes may be made to an approved hard hat or its suspension system. It is the responsibility of the wearer to assure that the hard hat has not been tampered with or altered.

Before wearing your hard hat, you should inspect it—each and every time you wear it. If there are any cracks, dents, or deep scratches in the shell, or if any webbing straps are worn or torn, you should get a new hard hat. Paints, solvents, and other chemicals may have an adverse effect on the shell and should not be used or allowed to come in contact with hard hats.

The bill of the hard hat is to be worn facing forward. This will provide added protection to your face and eyes. You should also wear the hard hat slightly raised off the head. This way, if something does fall on your head, the hard hat will cushion the effect, acting as a shock absorber. If your hair is long, tuck it up under the hat or tie your hair back out of the way.

Safety glasses, goggles, or a face shield, as shown in *Figure 4*, are used to protect the worker's eyes and face from flying chips of brick, dust, chemical vapor or splash, molten metal, and potentially injurious light radiation. Safety glasses or goggles should be worn whenever there is any chance of an eye injury, such as when you are chipping or cutting brick or block. A face shield should be used in situations when not only the eyes but the entire face must be protected, such as when grinding, buffing, handling corrosive chemicals, or using impact tools to chip or break brick or block.

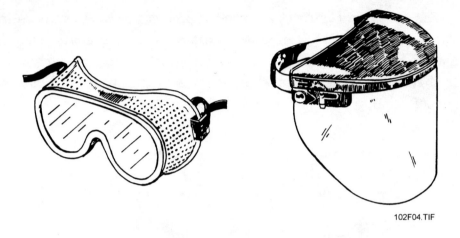

102F04.TIF

Figure 4. Goggles And Face Shield

All eye protection must meet the specifications in ANSI Z87.1. Different working conditions require different safety equipment. Chemical goggles, for example, should be worn when working with materials that are corrosive to skin tissue, such as grout. Chemical goggles will provide protection against chemical splashes, spray, impact of foreign particles, and exposure to dusty conditions. Chipper goggles are required for operations where there is danger of foreign particles striking from the sides, top, or bottom, such as when chipping, grinding, or splitting brick or block.

Welding goggles or hoods, as shown in *Figure 5*, are used to protect the worker's eyes when welding or cutting. They are required for use during all gouging or welding operations.

102F05.TIF

Figure 5. Welding Goggles

Before performing any work, make sure you have the correct eye protection for the job you will be performing. Careful consideration should be given to the kind of work you will be doing, the nature of the hazard, and the potential severity of injury. You should also inspect your eye and face equipment daily and clean it with a cleaning solution and tissue or cloth. Broken, dirty, pitted, or scratched eye and face protection equipment can be a source of reduced vision, is more likely to break, and does not protect you well. Inspect your eyewear frequently for any damage and replace it if necessary.

Hearing protectors, as shown in *Figure 6*, are used to protect the worker's ears from harsh or loud noise. Construction sites are full of noise. Most construction companies follow OSHA rules in deciding when ear protection must be used. Basically, though, ear protectors should be worn any time there is noise, such as when you are cutting brick or block with a power saw. To prevent ear infection, protectors should be cleaned regularly with soap and water.

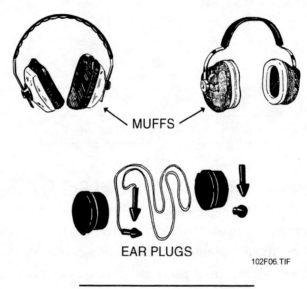

MUFFS

EAR PLUGS

102F06.TIF

Figure 6. Hearing Protectors

Lung protection devices are used to protect the worker's lungs from dust and dirt that may, over a prolonged exposure, cause silicosis. The devices should be worn whenever work is being done that puts fumes or particles into the air that may be dangerous to breathe for long periods of time. Breathers and masks, as shown in *Figure 7*, fit tightly over the nose and mouth and consist of a coarse filtering material that is easy and safe to breathe through. The device is held in place by two elastic bands. This type of breathing mask should be worn when cutting brick or block.

Dust or loose particles usually require only a breather, but noxious gases or fumes **require the better protection of a respirator**, as shown in *Figure 8*.

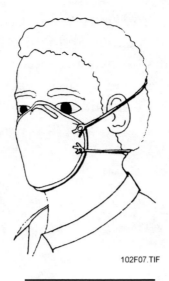

Figure 7. Dust Mask

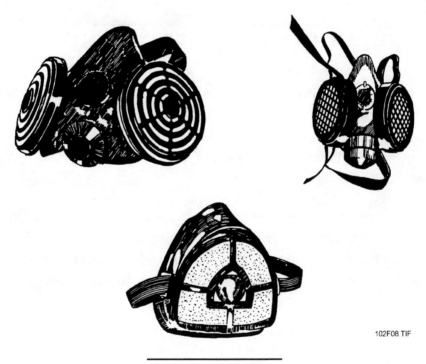

Figure 8. Respirators

Federal law dictates which type of respirator should be used to protect workers from different types of hazards. Before using a respirator, make sure you are trained in the proper procedures for using and maintaining a respirator. You should also check the respirator for damage and for a proper fit. And make sure you use the appropriate respirator for your specific situation and for the contaminant you will be exposed to. When you are finished using the respirator, make sure you clean it after each day of use, or more often as necessary. Respirators that are used by more than one person should be cleaned and disinfected after each use.

Leather and rubber gloves, as shown in *Figure 9*, are used to protect the worker's hands from **caustic** or **toxic** materials. Prolonged exposure to fresh mortar, for example, can cause skin irritation and possible further injury to existing cuts. That is why it is important to wear rubber gloves whenever handling caustic or toxic material, and leather gloves otherwise. When gloves become worn or torn, they should be replaced and/or disposed of properly.

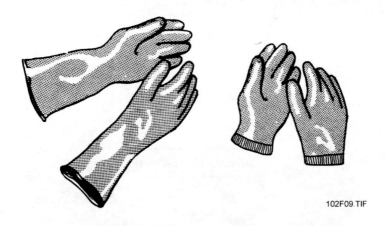

102F09.TIF

Figure 9. Rubber And Work Gloves

Fall protection equipment, as shown in *Figure 10*, is used to protect the worker from injuries that are caused by falls. It should be used when working on scaffolds, high-rise buildings, roofs, and other above-ground situations, as well as when working near a deep hole or a large opening in a floor, or when working above protruding rebar. Fall protection equipment includes a lanyard and body harness.

A body harness goes around legs and shoulders with strapping across the chest and back. It has a D-ring that is used to attach a lanyard. The lanyard is then attached to a strong anchor point that is capable of holding more than 5,400 pounds without failure. The lanyard should be long enough to let you work but also short enough to limit a fall. Some lanyards have built-in shock absorbers.

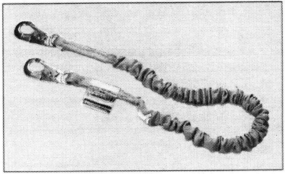

102F10.TIF

Figure 10. Fall Protection Equipment

Workers on suspended scaffolding may also use lifelines and rope grabs, as shown in *Figure 11*, with harnesses. The lifeline is secured to a point independent of the scaffold platform. The rope grab links the lifeline to the harness. The grab has a ratchet that locks in a fall.

Like all safety equipment, lanyards, lifelines, and safety harnesses should be carefully inspected before each use. Look for worn or frayed areas, check for metal fatigue, and do not use the equipment if you find any damage. It is important to use only equipment that meets or exceeds minimum OSHA standards.

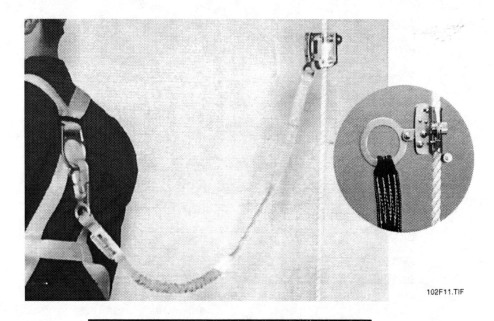

Figure 11. Detail Of Harness And Rope Grab

Safety shoes or boots, as shown in *Figure 12*, are used to protect the worker's feet from injury. The type of footwear you wear will be determined by the type of work you are doing.

Figure 12. Work Boots

Work boots should be equipped with sturdy heels so as to support the feet properly and reduce the chance of a turned ankle. Excessively thin soles on shoes should be avoided because they can cause sore feet and fatigue. Steel toes add extra protection from **dropped objects.**

2.5.0 WEARING SAFETY GEAR AND CLOTHING

In general, the employer is responsible to OSHA for making sure that all employees are wearing Appropriate Personal Protective Equipment whenever those employees are exposed to possible hazards to their safety. In turn, you are responsible for wearing the gear and clothing assigned to you.

Properly dressed masonry craftworkers are the workers who dress safely for the jobs they perform. It is important, for example, to take the following safety precautions when dressing for masonry work:

- Remove all jewelry, including wedding rings, bracelets, necklaces, and earrings. Jewelry can get caught on or in equipment, which could result in a lost finger, ear, or other appendage.

- Confine long hair in a ponytail or in your hard hat. Flying hair can obscure your view or get caught in machinery.

- Wear close-fitting clothing that is appropriate for the job. Clothing should be comfortable and should not interfere with the free movement of your body. Clothing or accessories that do not fit tightly, or that are too loose or torn, may get caught in tools, materials, or scaffolding.

- Wear face and eye protection as required or if there is a risk from flying particles, debris, or other hazards such as brick dust or chemicals.

- Wear hearing protection as required.

- Wear respiratory protection as required.

- Wear a long-sleeved shirt so as to provide extra protection for your skin.

- Protect any exposed skin by applying skin cream, body lotion, or petroleum jelly.

- Wear sturdy work boots or work shoes with thick soles. Never show up for work dressed in sneakers or gym shoes.

- Wear fall protection equipment as required.

Figure 13 shows a craftworker equipped with all Appropriate Personal Protective Equipment for most masonry jobs.

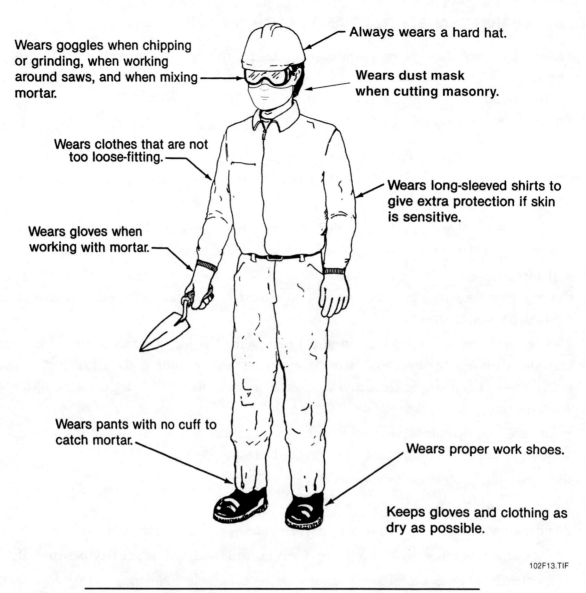

Wears goggles when chipping or grinding, when working around saws, and when mixing mortar.

Always wears a hard hat.

Wears dust mask when cutting masonry.

Wears clothes that are not too loose-fitting.

Wears long-sleeved shirts to give extra protection if skin is sensitive.

Wears gloves when working with mortar.

Wears pants with no cuff to catch mortar.

Wears proper work shoes.

Keeps gloves and clothing as dry as possible.

102F13.TIF

Figure 13. Wearing Appropriate Personal Protective Equipment

3.0.0 HAZARDS ON THE JOB

Construction sites may contain numerous hazards, as shown in *Figure 14*. You need to walk and work with all due respect for those hazards. The following list includes some of the hazardous conditions at a typical job site:

- improper ventilation
- inadequate lighting
- high noise levels
- slippery floors
- unmarked low ceilings

- excavations, holes, and open, unguarded spaces—including open, unbarricaded elevator shafts
- poorly constructed or poorly rigged scaffolds
- improperly stacked materials
- live wires, loose wires, and extension cords
- unsafe ladders
- unsafe crane operations
- water and mud
- unsafe storage of hazardous or flammable materials
- defective or unsafe tools and equipment
- poor housekeeping

102F14.TIF

Figure 14. Hazardous Construction Site

Your safety and that of your fellow workers should be a primary consideration in your work life. Some common-sense rules and ways of doing things can make the job site safer for everyone. These safety tips should be a part of your everyday thinking. Use your head to develop a positive safety attitude. It will keep you and the people you work with safe and sound.

3.1.0 HOUSEKEEPING

In order for a work site to be safe, each individual must act reasonably and follow established safety procedures. In addition, everyone must be constantly aware of conditions in the work area that may lead to injuries or illness or could endanger others.

The proper attitude toward injury and illness prevention is necessary on the part of all employees. It requires cooperation in all safety and health matters between workers. Only through such a cooperative effort can a safety program in the best interest of all be established and preserved. Safety is no accident. Think safety and the job will be safer.

One of the first steps to becoming safety conscious is being able to recognize safe on-the-job practices. You need to familiarize yourself with these practices.

One of the easiest ways to promote safety on the work site is to keep the work site clean and free of obstructions. This means planning your work to avoid unsafe conditions, maintaining a clean and efficient area while you work, and cleaning up after your work.

Another name for this is pride of workmanship. If you take pride in what you are doing, you simply will not allow trash to build up around you. You know that efficient work requires neatness and order, because you want to avoid always having to look through a pile of junk for something you need in a hurry.

The old idea of "a place for everything and everything in its place" may sound corny, but it is a good idea just the same. Part of this includes your scraps and garbage. The place for this is in the provided containers, and part of taking pride in your work is seeing that the garbage is always disposed of properly.

To establish a safe work site, you should make these your rules:

* Maintain the job site in a clean and orderly condition. This includes bending all exposed nails and staples so they do not injure anyone; securing any and all air, water, and electric lines so they do not become tripping hazards; and keeping all aisles and passageways clean and clear of obstructions.

* Encourage and enforce personal hygiene. This includes washing your hands after handling materials to remove any residue or hazardous substances that could lead to accidents or illnesses, and changing your clothes, gloves, or shoes if they become wet or get gasoline or solvent on them.

* Keep all work surfaces dry and slip resistant. If you do spill something, clean it up immediately.

* Store combustible scrap and waste safely, and dispose of it promptly.

* Remove accumulations of dirt in order to prevent dust from entering electrical enclosures or equipment.

Another way to promote safety is to maintain a good work attitude. This means conducting yourself in a professional manner and avoiding any horseplay, pranks, or fighting. Practical jokes can be fatal and should be avoided. It is also important that you take care not to show up at the work site under the influence of drugs or alcohol, or operate any equipment, tools, or vehicles when in an impaired state.

A good work attitude means that you will:

- Follow safety regulations and procedures at all times, including company and industry standards and federal, state, and local laws.
- Participate regularly in safety meetings.
- Get your share of sleep and good food; a good night's sleep and a full meal can help make you feel better and avoid accidents.
- Watch out for unsafe work methods or unsafe conditions and report them promptly to your supervisor.
- Report any injury or accident—serious or not—to your supervisor.
- Use appropriate safety warning devices to notify others of any danger zone.
- Use safety harnesses where necessary and be sure safety lines are tied off at an area that is free of the work platform.
- Assist other trades when necessary to maintain safe operations.
- Respect electricity by never using electrical equipment in areas of excessive moisture unless you have taken all safeguards, and by using only approved electric power tools and extensions that are grounded.
- Store materials and supplies safely in their proper places.
- Lift heavy objects with your legs—not with your back.
- Know the location of fire-fighting and first-aid equipment.
- Be conscious of others' safety and encourage others to do the same.

Just as there is a list of things to do to promote safety, there is also a list of things not to do.

- Do not attempt any work or repair for which you are not qualified.
- Do not perform hazardous work alone.
- Never place yourself in a dangerous position or allow others to work in a dangerous position. For example, never work or pass under suspended equipment.
- Do not tamper with, remove, or damage any protective devices or safety devices that are installed on tools or equipment.
- Do not leave electrical wires or equipment in water. Do not use wet extension handles around electrical equipment or power lines.

- Never enter any underground vault, manhole, silo, or enclosed area until you are certain that the air within contains no flammable, contaminated, or toxic vapor or gas.
- Never use compressed air to clean your clothes or yourself.

Keep in mind that the major goal of effective housekeeping is to prevent accidents. Do your part to help reduce the chances for falls, slips, fires, explosions, and falling objects.

3.2.0 TRENCHES AND EXCAVATIONS

Many construction jobs include the need to work in trenches and excavations. In order to prevent accidents and injury, safe work procedures must be followed by those workers performing this work and by any others in the area. Cave-ins and objects and tools falling into an excavation are constant threats to worker safety.

There are three safety points to remember when working in and around excavations:

- Protect the workers
- Protect the general public
- Protect nearby structures and utilities

Before digging, perform the following procedures:

- Check nearby structures, trees, rocks, etc., to see how they will affect the excavation.
- Check with local utility companies to find out if there are any underground utilities such as sewer, telephone, water, fuel, or electric lines.
- If utilities are there, indicate their location and depth with an above-ground stake so that these areas can be avoided.

Before entering the trench or excavation:

- Erect barricades, guardrails, or fences completely around the perimeter of the excavation, as shown in *Figure 15*.
- Remove all tools, materials, and loose dirt and rocks from the edge of the trench or excavation, as they can easily fall in and injure workers in the hole.
- Keep all heavy equipment and materials as far away from the excavation as possible; excessive weight near the edge of a trench could cause a cave-in.
- Test the atmosphere in the excavation for hazardous conditions. If necessary, use adequate precaution to prevent employee exposure to hazardous atmospheres and provide some type of ventilation.
- Check to make sure all personnel involved in the excavation procedures are properly dressed and are using prescribed safety equipment.

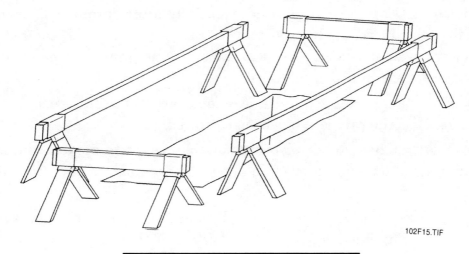

Figure 15. Barricade Around A Trench

- Provide some form of shoring or brace or sloping of the ground to prevent the excavation from caving in.
- Provide steps or ladders every 25 feet when the trench or excavation is 4 feet deep or more, so that in case of emergency, workers can leave the trench quickly.

When working in a trench or excavation, follow these precautions:

- Check the excavating procedures to make sure there are no hazards or neglected safety precautions.
- Shore, brace, or drive piling, as shown in *Figure 16*, if vibrations or blasting are prevalent near the excavation.

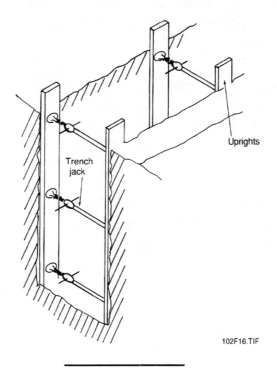

Figure 16. Shoring

- Inspect the shoring systems daily and especially after rainstorms or any change in conditions that may increase the possibility of cave-ins or slides. If signs of possible cave-ins or slides are evident, stop all work in the excavation until the necessary precautions have been taken to safeguard employees.

- Stop all work if dangerous ground movements are apparent, such as settlement or tension cracks, until the problem has been corrected.

- Provide diversion ditches, dikes, or other suitable means to prevent surface water from entering an excavation and to provide adequate drainage of the area adjacent to the excavation.

- Guard against an unstable excavation bottom, such as when working below the waterline, by driving sheeting below the bottom of the excavation to add to the soil stability. Water causes soil erosion and softening and should not be allowed to accumulate in a trench or excavation.

- Always walk around a trench. Never jump over or straddle a trench. You could lose your footing and fall in, or your weight could cause a cave-in.

- Always use a ladder to get into and out of the trench. Never jump into a trench.

- Wear a reflectorized warning vest if exposed to vehicular traffic.

- Stand away from any vehicle being loaded or unloaded in order to avoid being struck by any spillage or falling materials.

- Do not work in areas where there is accumulated water unless adequate precautions have been taken.

As soon as work is completed, the trench should be backfilled as the shoring is dismantled. After the trench has been cleared, workers should remove the shoring from the bottom up, taking care to release jacks or braces slowly. In unstable soil, ropes should be used to pull out the jacks or braces from above.

3.3.0   WORKING IN CONFINED SPACES

Construction and maintenance work is not always done in the great outdoors. Much of it takes place in **confined spaces** where the atmosphere is not easily ventilated, such as a kiln, manhole, or pressurized vessel.

OSHA defines a confined space as "any space having a limited means of egress, which is subject to the accumulation of toxic or flammable contaminants or has an oxygen deficient atmosphere. Confined or enclosed spaces include, but are not limited to, storage tanks, process vessels, bins, boilers, ventilation or exhaust ducts, sewers, underground utility vaults, tunnels, pipelines, and open top spaces more than 4 feet in depth such as pits, tubs, vaults, and vessels."

Many confined spaces normally contain hazardous gases and/or fluids. In addition, the work you are doing may introduce hazardous fumes into the space. Welding is an example of this. To ensure safety, special precautions are needed before entering a confined space and while working in a confined space.

When your work involves a confined space, you should always follow your employer's prescribed procedures. For example, confined space procedures may include receiving clearance from a safety representative before starting work and any time the job conditions change. In some instances, wearing the proper safety gear and clothing may not be enough to protect you from the hazards of the work site. For example, in cold weather, portable heaters may be set up in rooms in order to keep workers and materials warm. However, if proper ventilation is not provided, these heaters can eat up the oxygen in the air and produce carbon monoxide.

Odorless and tasteless, carbon monoxide can seep into your lungs, causing nausea, vomiting, headaches, and, if not caught in time, death. One way to prevent this gas buildup is to make sure all heaters are properly vented to the outside. However, even proper venting can sometimes cause problems. And because you cannot see, smell, or taste carbon monoxide, you may not be aware of its presence until it is too late.

To ensure that you become aware of deadly levels of carbon monoxide before they can cause you undue harm, you should use a carbon monoxide meter similar to the one shown in *Figure 17*. This type of meter will alert you when you are approaching dangerous levels of carbon monoxide so that you can leave the area before you are overtaken by the gas.

102F17.TIF

Figure 17. Carbon Monoxide Meter

Choose a meter that best fits your needs and learn how to operate it effectively. You should also calibrate it on a regular basis to make sure it is operating correctly. Remember, your life and others' lives could depend on how well you know how to operate a carbon monoxide meter.

Here are some additional procedures that should be followed when working in a confined space:

- Follow your company's prescribed entry procedure.

- Make sure you have a permit allowing entry to the confined space.

- Make sure all pipelines have been disconnected or blanked off. This will prevent other liquids, gases, or solids from inadvertently entering the confined space while you are in it.

- Make sure the atmosphere has been tested before entering the confined space and that you are wearing any and all required safety gear. Air sample readings should have been taken for levels of oxygen and explosive gases before you enter the area.

- Work with a team or standby person.

- Stay in voice or visual contact with someone outside the confined area. If you are the attendant or fire monitor, stay alert.

- Know your company's emergency procedures for confined space work.

- Use the required respiratory protection equipment.

- Use only approved electrical appliances, lights, extension cords, and tools.

- Never enter a confined space in an emergency unless you are equipped with the proper rescue equipment and there is someone else to stand by. Notify the proper personnel.

- Use safety harness equipment.

3.4.0 CONTROLLED ACCESS ZONES

A controlled access zone is a work area that is designated and clearly marked in which certain types of work such as overhand bricklaying and leading edge work may take place without the use of conventional fall protection systems such as guardrails, personal arrest systems, or safety nets. Controlled access zones are created to limit entrance to these areas so that only authorized workers may enter them. This helps protect your fellow workers.

Controlled access zones must be defined by a control line or by any other means that restrict access. Control lines should consist of ropes, wires, tapes, or equivalent materials, and supporting stanchions. OSHA requirements for erecting control lines to control access to areas where leading edge operations or overhand bricklaying or related work is being performed are provided in Section 1926.502 of OSHA Standards 1926. Where there are no guardrails, masons are the only workers allowed in controlled access zones.

On floors and roofs where guardrail systems are not in place prior to the beginning of overhand bricklaying operations, controlled access zones should be enlarged as necessary to enclose all points of access, material handling areas, and storage areas. On floors and roofs where guardrail systems are in place, but need to be removed to allow overhand bricklaying work or leading edge work to take place, only that portion of the guardrail necessary to accomplish that day's work shall be removed.

3.5.0 FALLING OBJECTS

The old rule says that "what goes up, must come down." Follow these guidelines to make sure that if something does come down, you are going to live to tell about it.

- Always wear a hard hat.

- Keep openings in floors covered. When guardrail systems are used to prevent materials from falling from one level to another, any openings must be small enough to prevent the passage of potential falling objects.

- Do not store materials other than masonry and mortar within 4 feet (1.2 m) of the working edges of a guardrail system.

- Keep the working area clear by removing excess mortar, broken or scattered masonry units, and all other materials and debris on a regular basis.

- Be very careful around operating cranes. Stay clear of the working area.

- Never work or walk under loads that are being hoisted by a crane.

- Learn the basic hoisting signals for cranes. Be sure to learn the stop signal and the emergency stop signal, as shown in *Figure 18*.

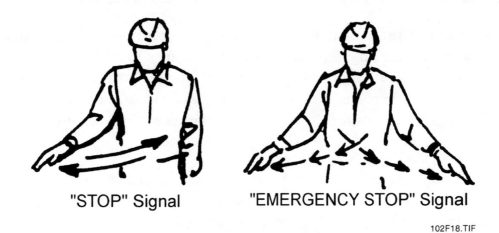

"STOP" Signal   "EMERGENCY STOP" Signal

102F18.TIF

Figure 18. Hoisting Signals

- Erect toeboards or guardrail systems to protect yourself from objects falling from higher levels. Toeboards should be erected along the edges of the overhead walking/working surface for a distance that is sufficient to protect the workers below.

- Erect paneling or screening from the walking/working surface or toeboard to the top of a guardrail system's top rail or midrail if tools, equipment, or materials are piled higher than the top edge of the toeboard.

- Raise or lower tools or materials with a rope and bucket or other lifting device. Never throw tools or materials.

- Never put tools or materials down on ladders or in other places where they can fall and injure people below. Before moving a ladder, make sure there are no tools left on it.

3.6.0   FIRE PREVENTION

Another danger of working in the masonry trade is exposure to potential fire hazards. Here are some basic safety guidelines for fire prevention:

- Always work in a well-ventilated area, especially when using flammable materials such as shellac, lacquer, paint stripper, and construction adhesives.

- Never smoke or light matches when working with flammable materials.

- Keep oily rags in metal waste cans.

- Store combustible substances in approved containers.

- Never use a pail of water on a liquid or electrical fire. Water will spread the fire or cause electrical shock.

- Make sure all fire extinguishers are fully charged. Never remove the tag from an extinguisher—it indicates the last date it was serviced and inspected.

If there is a fire, it is important that you know the location of fire extinguishers, the appropriate type to use for different kinds of fires, and how to use them. Fire extinguishers are labeled according to the class or classes of fires on which they may be used. The class letters are prominently displayed in distinctive shapes and colors. Most new extinguishers also contain pictorial labels depicting the class or classes of fires on which the extinguisher may or may not be used.

It is critical that masonry craftworkers be able to quickly determine the source of a fire in order to select the correct fire extinguisher to fight it. Knowing which fire extinguisher should be used on which class or classes of fire can help save lives.

- *Class A fire extinguishers* are designed to be used to control fires that occur in ordinary combustible materials such as scrap lumber, packaging, insulation, rags, rubber, and plastics.
- *Class B fire extinguishers* are designed to be used to control fires that occur with flammable liquids such as gasoline, lubricants, solvents, adhesives, oil, grease, paint, and thinners.
- *Class C fire extinguishers* are designed to be used to control fires that occur in or near electrical equipment such as motors, switchboards, and electrical wiring.
- *Class D fire extinguishers* are designed to be used to control fires that occur with combustible metals such as iron, magnesium, sodium, and potassium. They may be necessary when welding, cutting, and performing other operations that produce high temperatures.

Figure 19 provides a chart that you can use to recognize the different types of fires and extinguishers used to fight them.

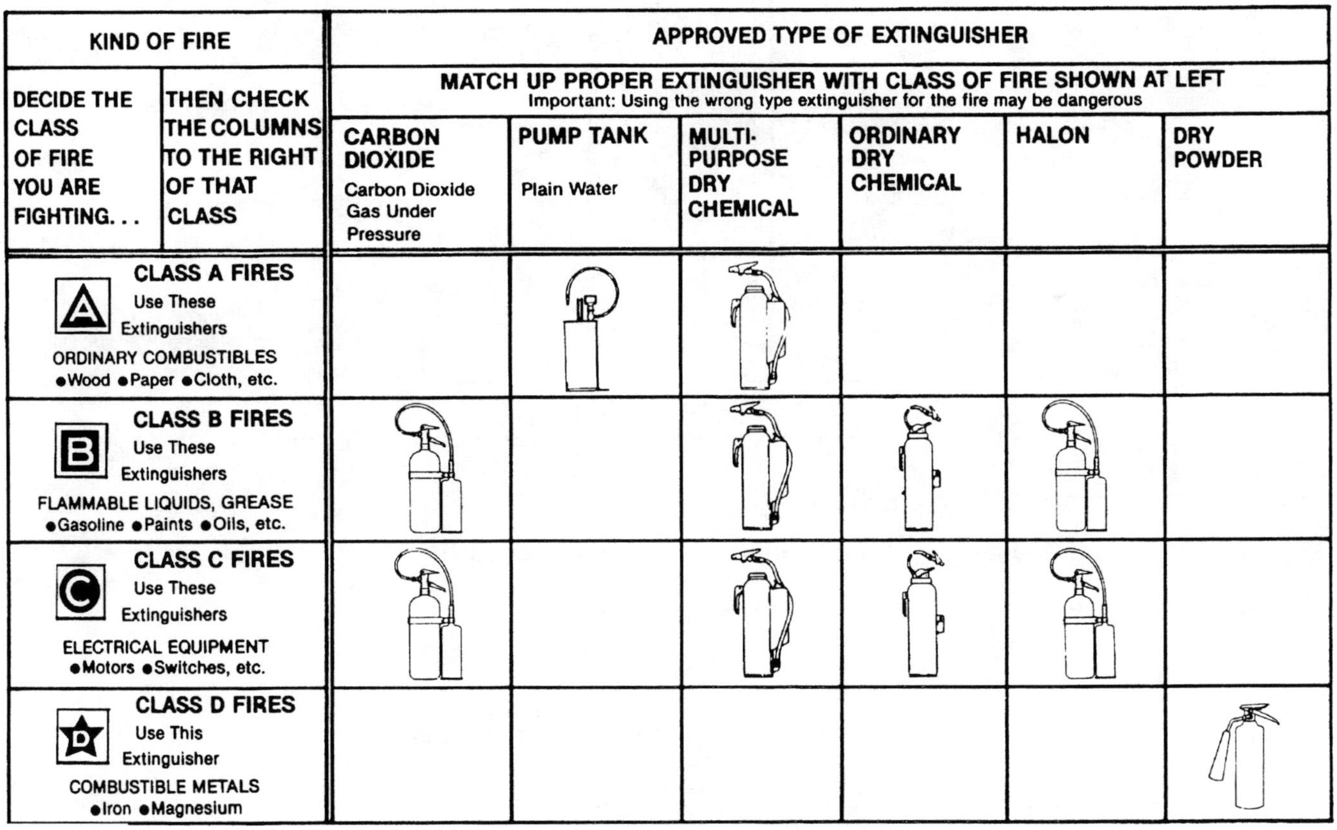

KIND OF FIRE		APPROVED TYPE OF EXTINGUISHER					
		MATCH UP PROPER EXTINGUISHER WITH CLASS OF FIRE SHOWN AT LEFT Important: Using the wrong type extinguisher for the fire may be dangerous					
DECIDE THE CLASS OF FIRE YOU ARE FIGHTING...	THEN CHECK THE COLUMNS TO THE RIGHT OF THAT CLASS	CARBON DIOXIDE Carbon Dioxide Gas Under Pressure	PUMP TANK Plain Water	MULTI-PURPOSE DRY CHEMICAL	ORDINARY DRY CHEMICAL	HALON	DRY POWDER
CLASS A FIRES Use These Extinguishers ORDINARY COMBUSTIBLES ●Wood ●Paper ●Cloth, etc.			⬤	⬤			
CLASS B FIRES Use These Extinguishers FLAMMABLE LIQUIDS, GREASE ●Gasoline ●Paints ●Oils, etc.		⬤		⬤	⬤	⬤	
CLASS C FIRES Use These Extinguishers ELECTRICAL EQUIPMENT ●Motors ●Switches, etc.		⬤		⬤	⬤	⬤	
CLASS D FIRES Use This Extinguisher COMBUSTIBLE METALS ●Iron ●Magnesium							⬤

102F19.TIF

Figure 19. Types Of Fire Extinguishers

3.7.0   ELECTRICAL SAFETY

Another potential danger for masonry craftworkers is electricity. OSHA is very specific about keeping the workplace safe from electrical hazards. There are many things you can do to reduce the chance of a fatal electrical accident. Here are the basic safety guidelines for working with or near electricity:

- Always inspect electrical power tools before each use.
- Make sure all panels, switches, outlets, and plugs are grounded. Cords for portable power tools must be the 3-wire type (see *Figure 20*) and they must be properly connected. The 3-wire system is one of the most common safety grounding systems used to protect you from accidental electrical shock. The third wire is connected to ground. Should the insulation in a tool fail, the current will pass to ground through the third wire—and not through your body.

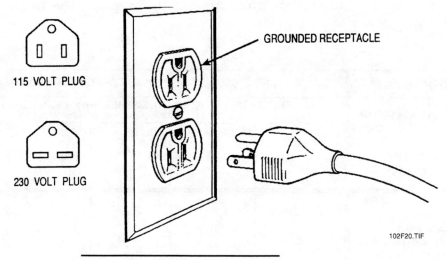

GROUNDED RECEPTACLE

115 VOLT PLUG

230 VOLT PLUG

102F20.TIF

Figure 20.   Three-Wire System

- Do not use worn or frayed cables.
- Make sure all double-insulated tools are Underwriter Laboratory (UL) approved.
- Make sure all bulbs have guards (see *Figure 21*).

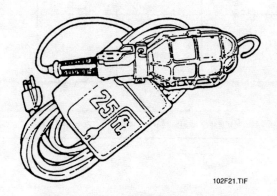

102F21.TIF

Figure 21.   Bulb With Guard

- Do not hang temporary lights by their cords unless they were designed for it.
- Never use damaged cords.
- Never use bare electrical wire.
- Never use metal ladders near electrical installations.
- Never wear a metal hard hat.
- Never operate any device that has a danger tag/lockout device attached.

With the wide use of portable power tools on construction sites, it is often necessary to use flexible cords. However, cords, cord connectors, receptacles, and cord- and plug-connected equipment can cause hazards, especially if they are improperly used and maintained.

Flexible cords are generally more vulnerable to damage than is fixed wiring. To prevent tension at joints and terminal screws, flexible cords must be connected to devices and to fittings.

Because the cord is exposed, flexible, and unsecured, joints and terminals also become more vulnerable. Flexible cord conductors are finely stranded for flexibility, but the strands of one conductor may loosen from under terminal screws and touch another conductor, especially if the cord is subjected to stress or strain.

A flexible cord can be damaged by activities on the job, by door or window edges, by staples or fastenings, by abrasion from adjacent materials, or simply by aging. If the electrical conductors become exposed, there is a danger of shocks, burns, or fire.

When a cord connector is wet, hazardous leakage can occur to the equipment grounding conductor and to humans who pick up that connector if they also provide a path to ground. Such leakage is not limited to the face of the connector but also develops at any wetted portion of it.

When the leakage current of tools is below 1 ampere and the grounding conductor has a low resistance, no shock should be perceived. However, should the resistance of the equipment grounding conductor increase, the current through the body will also increase. Thus, if the resistance of the equipment grounding conductor is significantly greater than 1 ohm, tools with even small leakages become hazardous.

One way to reduce electrical hazards on construction sites is to use ground fault circuit interrupters (GFCIs), as shown in *Figure 22*. Tripping of GFCIs, or interruption of current flow, is sometimes caused by wet connectors and tools. It is good practice to limit exposure of connectors and tools to excessive moisture by using watertight or sealable connectors. Providing more GFCIs or shorter circuits can prevent tripping caused by the cumulative leakage from several tools or by leakages from extremely long circuits.

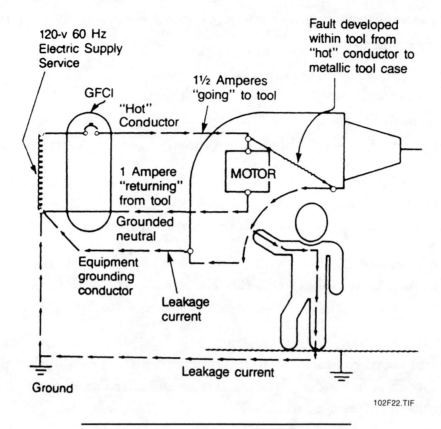

120-v 60 Hz
Electric Supply
Service

GFCI

"Hot"
Conductor

1½ Amperes
"going" to tool

Fault developed
within tool from
"hot" conductor to
metallic tool case

1 Ampere
"returning"
from tool

MOTOR

Grounded
neutral

Equipment
grounding
conductor

Leakage
current

Leakage current

Ground

102F22.TIF

Figure 22. Ground Fault Circuit Interrupter

The GFCI is a fast-acting circuit breaker that senses small imbalances in the circuit caused by current leakage to ground and, in a fraction of a second, shuts off the electricity. However, the GFCI will not protect the employee from line-to-line contact hazards (such as a person holding two hot wires or a hot and a neutral wire in each hand). It does provide protection against the most common form of electrical shock hazard—the ground fault. It also provides protection against fires, overheating, and the destruction of insulation on wiring.

3.8.0   MORTAR AND CONCRETE SAFETY

Another hazard encountered by masonry craftworkers is exposure to mortar, grout, and concrete. These cement based materials have ingredients that can hurt your eyes or skin. The basic ingredient of mortar, portland cement, is **alkaline** in nature and therefore caustic. It is also **hygroscopic**, which means that it tends to absorb moisture from your skin. Prolonged contact between the fresh mix and skin can cause skin irritation and chemical burns to hands, feet, and exposed skin areas. It can also saturate a worker's clothes and transmit alkaline or hygroscopic effects to the skin. In addition, the sand contained in fresh mortar can cause skin abrasions through prolonged contact.

Needless injury can be avoided by taking the following precautions:

- Keep your thumb on the ferrule of the trowel away from the mortar.
- Keep cement products off your skin at all times by wearing the proper protective clothing, including boots, gloves, and clothing with snug wristbands, ankle bands, and neckband. Make sure they are all in good condition.
- Prevent your skin from rubbing against cement products. Rubbing increases the chance of serious injury. If your skin does come in contact with any cement products, wash your skin promptly. If a reaction persists, seek medical attention.
- Wash thoroughly to prevent skin damage from cement dust.
- Keep cement products out of your eyes by wearing safety glasses when mixing mortar or pumping grout. If any cement or cement mixtures get in your eye, flush it immediately and repeatedly with water and consult a physician promptly.
- Rinse off any clothing that becomes saturated from contact with fresh mortar. A prompt rinse with clean water will prevent continued contact with skin surfaces.
- Be alert—watch for trucks backing into position and overhead equipment delivering materials. Listen for the alarms or warning bells on mixers, pavers, and ready-mix trucks.
- Never put your hands or arm, or any tool, into rotating mixers.
- Use good work practices to reduce dust in the air when handling mortar and lime. For example, do not shake out mortar bags unnecessarily and do not use compressed air to blow mortar dust off clothing or a work surface. Stand upwind when dumping mortar bags.
- Be certain adequate ventilation is provided when using epoxy resins, organic solvents, brick cleaners, and other toxic substances.
- Never use solvents to clean skin.
- Immediately remove epoxy, solvents, and other toxic substances from skin. Use appropriate cleansing agents to remove these substances.

4.0.0 MATERIALS HANDLING

According to OSHA, there are an estimated 575,000 existing chemical products, and hundreds of new ones are being introduced annually. This poses a serious problem for workers. Exposure to chemicals can cause or contribute to many serious health defects, such as heart ailments, kidney and lung damage, sterility, cancer, burns, and rashes. Some chemicals may also be safety hazards and have the potential to cause fires, explosions, and other serious accidents.

OSHA has a rule called Hazard Communication, whose basic goal is to reduce the incidence of chemical source illness and injuries. It requires informing employers and employees about work hazards and how to protect themselves. As part of this standard, chemical manufacturers, importers, and distributors must be sure that all containers of hazardous chemicals arriving at the workplace are labeled, tagged, or marked with the chemical's identity, appropriate hazard warnings, and the name and address of the manufacturer or other responsible party.

Your employer has a responsibility to ensure that each container in the workplace is labeled, tagged, or marked with the identity of hazardous chemicals contained therein, and shows hazard warnings appropriate for employee protection. Your responsibility is to adhere to the warnings on these labels and use the materials appropriately.

Hazard communication and emergency response procedures are important parts of a project's safety and health program. It is also part of OSHA requirements that you and OSHA have access to this information. Everyone has a right to information on hazards, including chemical hazards.

You can get information on hazardous chemicals, substances, and products from container labels and from the instruction of your supervisor. In addition to product labels, manufacturers provide Materials Safety Data Sheets (MSDS) which provide added detail on the hazards of a product, how to control the hazards, what to do in the event of an emergency, and how to protect yourself. MSDS are collected and maintained for your use, reference, and easy access. They are summarized in an inventory list of hazardous substances for your easy reference. Other essential information includes a readily available copy of the OSHA Hazard Communication Standard and your company's Hazard Communication Program and Emergency Response Procedures.

The MSDS program was developed to provide both the employer and employee with information necessary for working safely with hazardous chemicals. It provides the following information:

- Health hazards, such as hazard rating, type of hazard, symptoms of exposure, and effects of exposure
- Steps for emergency first aid
- Fire, explosion, and reactivity data, such as extinguishing agents and firefighting methods, products formed by fire or excessive heat, conditions to avoid, stability, incompatible materials and reaction, products of decomposition, and hazardous polymerization
- Types of protection equipment, including personal protection, ventilation, and additional protective equipment
- Handling and storage methods, leakage and spill control, and waste disposal
- Instructions for medical management, such as general instructions, personal hygiene, medical records, and exposure limits
- Physical and chemical properties data, such as physical state, color, chemical formula, and specific chemical composition
- Other safety information

A sample MSDS is depicted in *Figure 23*.

MATERIAL SAFETY DATA SHEET	SUBSTANCE (Chemical Name) Acetylene (Ethine, Ethyne)	NO. BG-2

PRODUCT NAME, NUMBER, SYNONYM CAS #74-86-2	COMMON OR TRADE NAME Acetylene	
MANUFACTURER'S NAME AND ADDRESS	TELEPHONE NUMBER – –	

HEALTH HAZARDS

HAZARD RATING	(X) DANGER	( ) WARNING	( ) CAUTION

TYPE OF HAZARD   Asphyxiant Gas, Flammable, Explosion

SYMPTOMS OF EXPOSURE
  Headache, breathlessness, weakness, dizziness, loss of consciousness

EFFECTS OF EXPOSURE   At concentration of 20% or more acetylene-headache, breathlessness, weakness. At concentration of 40% or more acetylene-collapse, loss of consciousness, fatality.

EMERGENCY FIRST AID

Remove exposed person at once to uncontaminated area, keep warm and quiet, administer oxygen if loss of consciousness has occurred, begin artificial respiration if breathing has stopped, call a physician at once.

FIRE, EXPLOSION, AND REACTIVITY DATA

EXTINGUISHING AGENTS AND FIRE FIGHTING METHODS   Use Class C or Class ABC Carbon Dioxide, dry chemical, or multi-purpose dry chemical extinguisher. Use water in spray form to keep exposed facilities cool. Attempt to stop *

FLASH POINT		FLAMMABLE OR EXPLOSIVE LIMIT	
OPEN CUP °C CLOSED CUP 17.8 °C		LOWER 2.5 % UPPER 82 %	
IGNITION TEMPERATURE °C		AUTO-IGNITION TEMPERATURE 300 °C	

PRODUCTS FORMED BY FIRE OR EXCESSIVE HEAT
                    Hydrogen, carbon

CONDITIONS TO AVOID   Avoid excessive build up of acetylene which may lead to flash fire or explosion. Avoid heat and open flames. Remove sources of accidental ignition.

STABILITY   ( ) Stable   (X) Unstable—Explain Conditions   Do not use, generate or distribute through hose or pipe, acetylene at pressures in excess-15 psig.

INCOMPATIBLE MATERIALS AND REACTIONS   Forms spontaneously explosive compounds with copper, silver, and mercury. Will react explosively with chlorine, **

PRODUCTS OF DECOMPOSITION
                    Hydrogen, carbon

HAZARDOUS POLYMERIZATION   (X) Will not occur   ( ) May occur—Explain Reaction and Products

PROTECTION EQUIPMENT

PERSONAL PROTECTION   Respiratory protection-self-contained breathing apparatus and positive pressure hose mask, or air-line mask. Protective clothing—leather gloves for cylinder handling.

VENTILATION   Local exhaust—sufficient to maintain concentration below 2.5% lower explosive limit.

ADDITIONAL PROTECTIVE EQUIPMENT   For welding operations involving acetylene—use appropriate protective equipment.

102F23.TIF

Figure 23. Sample MSDS

Before using any hazardous material, you should review these procedures:

- Verify that all containers are clearly labeled as to their contents, with the appropriate hazardous warning and the name and address of the manufacturer listed. No container of hazardous materials should be accepted without proper labeling in place.

- Know the hazardous properties of the chemicals you work with. Read the labels on all containers and follow the manufacturer's instructions for using them.

- Know emergency response procedures and practices. For example, you should know your company's emergency procedures regarding whom to call and when.

- Do not transfer hazardous materials to temporary containers other than for immediate use. At the close of work, leftover amounts are to be returned to the original container or turned over to the immediate supervisor for proper disposal.

- Store chemicals in a safe manner and according to the manufacturer's recommendations. Keep chemical containers closed when not in use.

- Know where your site's hazard communication program records for all hazardous materials or substances are located. Your site should have an MSDS for all hazardous materials or substances.

- Know the first-aid treatment for any chemical you work with and be prepared to carry out that treatment immediately if need be.

- At regular intervals, inspect containers and pipelines that hold corrosive materials. Report leaks immediately.

- When using corrosive materials, know where the closest shower and eye fountain are located. Keep the shower and eye fountain unobstructed and in good working condition.

- Use a gas mask, goggles, gloves, and protective equipment as required.

4.1.0 TOXIC SUBSTANCES

There are a number of toxic substances a masonry craftworker might encounter on a job. These include acetone, arsenic, asbestos, benzene, cadmium, carbon monoxide, carbon tetrachloride, chlorine, chromates, coal dust, cobalt, DDT, epoxies, fiberglass, hydrochloric acid, lead compounds, nickel compounds, ozone, petroleum oils, burning plastic film and polyurethane foam, and sulfuric acid.

Toxic substances can be in solid, liquid, or gas form. And, when changed from their common form, they may still be toxic. For example, solids such as cement and asbestos are just as toxic when inhaled as airborne dusts. Solids are the least likely form to cause poisoning, but when they are changed into another form, they can be deadly, such as when welding rods decompose into fumes and gases, when burning plastics give off deadly gases, and when wood, concrete, and asbestos produce harmful dust.

Dust, gases, fumes, vapors, and mists can all be very dangerous when they enter the body through breathing, through the skin, or through open sores. Toxic substances can be inhaled, absorbed through the skin, ingested, or contracted through open cuts and sores. Open cuts and sores come in direct contact with toxic substances when they are touched by contaminated hands or when dust, fumes, and mists settle on them from the air.

Toxic substances such as carbon monoxide, airborne pesticides, and dusts may be inhaled directly from the air when you are breathing. Smokers can also inhale gases, dust, fumes, and mists through usual smoking, or when smoking cigarettes, pipes, or cigars that have been contaminated by direct contact with toxic substances. Some lead compounds, pesticides, and solvents can be absorbed directly through the skin. Food and drink may be contaminated from contact with toxic substances on the hands, or from dust, fumes, and mists that settle on them from the air.

Some chemicals may only affect the bodily areas they contact, whereas other chemicals may penetrate into one area of the body and then affect other bodily systems. Still others may seem to cause little bodily damage after exposure. However, without proper care, severe damage or death may be the result. That is why it is so important to know the substances you work with and what their properties are.

Some toxic substances may cause mild irritation, such as skin rashes, headaches, burning eyes, or coughing. Others may cause severe damage to any or all of the body's systems, such as shortness of breath, coma, shock, hemorrhage, or allergies. Still others can cause terminal conditions or diseases such as cancer, black lung, suffocation, or emphysema. It is your job to know what you are dealing with and take the necessary precautions to protect yourself.

Epoxy grout is a toxic substance that is used often in masonry work. It contains silica, which is very dangerous if inhaled. It is also an irritant to skin and a hazard to eyes. The hardener in epoxy can also cause severe chemical burns. When working with epoxy, it is important to wear safety goggles, chemical-resistant gloves, full-coverage clothing, and a respirator. In case of skin contact or respiratory problems, follow recommended first-aid procedures. For eye contact, prompt medical attention is essential.

Oxygen and acetylene cylinders can also be dangerous. To prevent accidents, it is important to secure them against rolling or tipping. You should also carefully handle any tanks or containers that contain explosive vapor or liquid, and never expose these containers to open flame or sparks.

Most of the internal organs and the external organs—the skin, eyes, nose, and throat—can be damaged by toxic materials. A lack of knowledge of toxic substances could be very dangerous. By paying attention, you can save yourself and your fellow workers from very painful and very dangerous situations.

4.2.0 FLAMMABLE LIQUID SAFETY

Flammable liquids require additional safety precautions:

- Carefully read labels on all flammable liquids, and use flammable liquids only in open, well-ventilated areas.
- Do not inhale or ignite fumes from flammable liquids.
- Be sure that all flammable liquids are marked correctly for storage.
- Store all flammable liquids properly and only in approved safety containers.
- Store oily rags and flammable materials in metal containers with self-sealing lids.
- When clothing is soaked by a flammable liquid, immediately change clothing and cleanse the body with an appropriate cleaner.
- Use flammable liquids only for their intended purposes; for example, never use gasoline as a cleaner.
- Do not use flammable liquids near fire or flame.
- Always be aware of the location of an appropriate fire extinguisher.

4.3.0 MATERIALS HANDLING

Approximately 25 percent of all occupational injuries occur when handling or moving construction materials. Strains, sprains, fractures, and crushing injuries can be minimized with a knowledge of safe lifting and handling procedures and good body mechanics. General pointers for you to keep in mind include the following:

- Wear steel-toe safety shoes.
- Keep floors free of water, grease, and other slippery substances so as to prevent falls.
- Inspect materials for grease, slivers, and rough or sharp edges.
- Determine the weight of the load before applying force to move it.
- Know your own limits for how much you can lift.
- Be sure that your intended pathway is free from obstacles.
- Make sure your hands are free of oil and grease.
- Take a firm grip on the object before you move it, being careful to keep your fingers from being pinched.
- When the load is too large or heavy, get help, or, if possible, simply reduce the load and make more trips.
- When stacking materials, be sure to follow OSHA regulations as to the height, shape, and stability of the pile.

Materials in the general working area and on the stockpile should be stacked safely. Masonry units that are not stacked properly and secured in some way are very likely to fall, which could cause injury to you or a fellow worker.

4.3.1 Materials Stockpiling And Storage

OSHA has several guidelines regarding the stockpiling and handling of materials.

- All materials stored in tiers should be stacked, racked, blocked, interlocked, or otherwise secured to prevent sliding, falling, or collapse.

- Maximum safe load limits of floors within buildings and structures should be posted in all storage areas, and maximum safe loads should not be exceeded.

- Aisles and passageways should be kept clear to provide for the free and safe movement of material handling equipment or employees.

- Materials stored inside buildings under construction should not be placed within 6 feet of any hoistway or inside floor openings, nor within 10 feet of an exterior wall that does not extend above the top of the material stored.

- Each employee required to work on stored material in silos, hoppers, tanks, and similar storage areas should be equipped with personal fall arrest equipment.

- Noncompatible materials should be segregated in storage.

- Bagged materials should be stacked by stepping back the layers and cross-keying the bags at least every 10 bags high.

- Materials should not be stored on scaffolds or runways in excess of supplies needed for immediate operations.

- Stockpiles of palletized brick should not be more than 7 feet in height. When a loose brick stockpile reaches a height of 4 feet, it should be tapered back 2 inches in every foot of height above the 4-foot level.

- When loose masonry blocks are stockpiled higher than 6 feet, the stack should be tapered back one-half block per tier above the 6-foot level.

4.3.2 Working Stacks

Masons use working stacks of brick at the wall, on the ground, or on the scaffold platform. Palletized brick is unbundled and moved to where the mason will need it. When stacking materials in working piles, keep the pile neat and vertically in line to eliminate the possibility of snagging your clothes. Keep the piles about 3 feet high so that it will not be hard for the mason to get at the brick. Bricks and other materials stacked too high not only pose a safety hazard but may drastically curtail production.

The most common way to stack bricks is to reverse the direction of every other course so that the stack is secure, as shown in *Figure 24*. Such a stack should be no more than 3 feet high and no closer than 2 feet from the wall.

102F24.TIF

Figure 24. Stack Of Bricks

Wider stacks can be made by alternating a pattern of eight bricks, as shown in *Figure 25*. The bricks in the second layer are turned 90 degrees to the bricks in the first layer. This kind of working stack should be no more than 3 feet high and no closer than 2 feet from the wall.

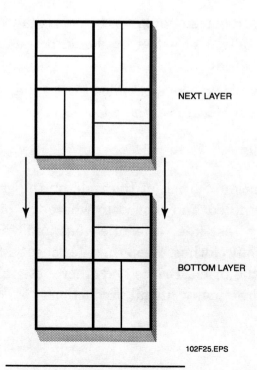

NEXT LAYER

BOTTOM LAYER

102F25.EPS

Figure 25. Stacking Pattern

Block is not put into working piles but lined up along the wall 2 feet away. Block is lined up on its plain end so that the ears are on top, ready to be buttered with mortar.

5.0.0   LADDER AND SCAFFOLDING SAFETY

More people are injured in slips and falls than in any other type of accident. Approximately one-third of all construction injuries and deaths involve falls from elevations. Many times, the most simple tasks are taken for granted, resulting in a slip or trip and—once "the point of no return" is reached—there is often a painful or disabling fall.

Falls are classified into two groups: falls from an elevation and falls on the same level. Because falls from an elevation are the more serious of the two types, scaffolding, work platforms, decking, and other staging is required to be fully enclosed by proper railings so as to limit falls to the same level, and reduce the severity of injury. This is not to say that falls on the same level cannot be extremely dangerous because a worker could fall and strike a sharp edge or be stabbed by some pointed object, but overall, falls from an elevation have more serious consequences.

Ideally, you want to prevent falls from happening. This involves building safe platforms enclosed by proper railings so that there is no opening to fall through from an elevated area, and providing a nonslip and level walking surface that has no holes or protrusions. In those instances when it is impossible to provide safe platforms and railings, you will need to protect yourself from falls by using fall protection equipment such as harnesses and lifelines, shock-absorbing lanyards, and safety nets.

The following safe practices can help prevent slips and falls:

- Wear safe, strong work boots that are in good repair.
- Watch where you step. Be sure your footing is secure.
- Do not allow yourself to get in an awkward position. Stay in control of your movements at all times.
- Maintain clean, smooth walking and working surfaces. Fill holes, ruts, and cracks. Clean up slippery material.
- Pick up litter. Do not allow tripping hazards to exist.
- Install cables, extension cords, and hoses so that they will not become tripping hazards.
- If you must climb to reach something, use a sound ladder that has been safely set in position and properly secured at both the top and bottom. Use only Type 1 industrial ladders.
- When climbing a ladder, always face the ladder and use both hands.
- When reaching from a ladder, keep your shoulder inside the vertical stringer. Do not overreach. Move the ladder.
- Walk, do not run.

OSHA recognizes that accidents involving falls are generally complex events frequently involving a variety of factors. Consequently, the standard for fall protection deals with both the human and the equipment related issues in protecting workers from fall hazards. For example, employers need to do the following tasks to ensure workplace safety:

- Assess the workplace to determine if the walking/working surfaces on which employees are to work have the strength and structural integrity to safely support workers.

- Prohibit employees from working on those surfaces until it has been determined that the surfaces have the requisite strength and structural integrity to support the workers.

- Select the appropriate fall protection system for the work operation if a fall hazard is present. For example, if an employee is exposed to falling 6 feet or more from an unprotected side or edge, the employer must select either a guardrail system, safety net system, or personal fall arrest system to protect the worker.

- Construct and install all safety systems properly.

- Supervise employees properly.

- Use safe work procedures.

- Train workers in the proper selection, use, and maintenance of fall protection systems.

5.1.0 LADDERS

Just as there are rules for working safely at a work site, there are also rules for working safely around construction equipment. Most construction sites, for example, are full of ladders and various types of scaffolding. In fact, the ladder is one of the most important tools on today's job site. When selecting a ladder for a job, it is important to choose one that will extend at least 36 inches above the eave or the roof line or landing. You should also always place the base of the ladder so that the distance between the feet and the wall is one-fourth of the height from the feet to the point where the ladder touches the wall, as illustrated in *Figure 26*.

If you set up a ladder in front of a door or entrance, make sure you lock the door or board it over so that you will not be knocked over in case someone tries to open the door. You should also watch out for overhead electrical lines and stay as far away from them as possible.

The following safety check should be performed before using a ladder:

- Check the specified work load of the ladder. Do not use a ladder if it has a smaller work load than what you need.

- Check the ladder for loose rungs, slivers, cracks, and splits in the wood, as well as loose nails, screws, or bolts.

- Check the ladder to make sure it has a nonslip base and a safety top.

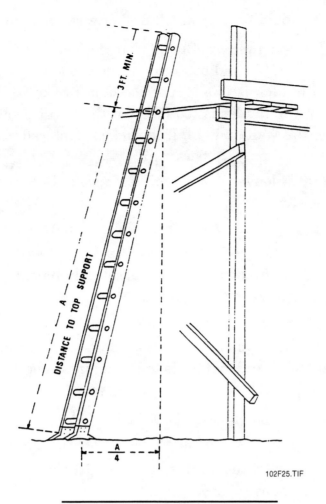

Figure 26. Safe Ladder Placement

- Check various parts of the ladder depending on what type you have. **Stepladders**, for example, should be checked for instability, which is often caused by loose or bent hinge spreaders, broken steps, or loose hinges. **Extension ladders** should be checked for defective or broken extension locks or rope deterioration. Wooden ladders should be checked to make sure they have not been painted. Paint could cover up a structural defect in a ladder.

If you find a defect in your ladder, do not just tag the defective part and use it anyway. Do not use it—period! Send any ladder with a defect back to the repair department, as per OSHA requirements. And, also per OSHA guidelines, do not use a metal ladder when performing any type of electrical work or whenever there is a possibility that you might come into contact with electrical conductors.

The following precautions should be kept in mind when setting up and using a ladder:

- Place the ladder in a stable manner. Stepladders, for example, should be opened fully on a solid, even surface. Portable and extension ladders should be placed on a solid base in such a way that the ladder is one-fourth the distance of its length from its vertical support. If the base of the ladder is too far from the vertical plane of its top support, it may slip farther from the wall and fall. If the ladder is placed too close to the wall, or the vertical plane of support, it may tip backward.

- Place the ladder so that it leaves 6 inches of clearance in back of the ladder and 30 inches of clearance in front of the ladder.

- Place the ladder so that it leans against a solid and immovable surface. Never place a ladder against a window, door, doorway, sash, loose or movable wall, or box.

- Do not use ladders during high wind. If you absolutely have to, make sure you lash the ladder securely in order to prevent slippage.

- Check to make sure the soles of your shoes are free of oil, mud, and grease.

- Keep both hands free so that you can hold the ladder securely while climbing. Use a rope to raise and lower any tools and materials that you might need.

- Never rest any tools or materials on the top of a ladder.

- Face the ladder whenever you climb up or down it.

- Climb or descend the ladder one rung at a time. Never run up or slide down a ladder.

- Move the ladder in line with the work to be done. Never lean sideways away from the ladder in order to reach the work area.

- Never stand on the top two rungs of a ladder.

- Use ladders for only short periods of off-ground work. If you must work from a ladder for extended periods, use a lifeline fastened to a safety harness.

- Lay the ladder down on the ground when you have finished using it, unless it is anchored securely at the top and bottom where it is used.

- Do not use ladders that are not equipped with safety shoes.

- Never overreach from a ladder.

- Never use ladders in a horizontal position.

- Do not climb a ladder that is occupied by someone else.

- Secure ladders at both the top and bottom to prevent displacement. Brace long ladders at intermediate points to prevent spring.

- Do not tie or fasten ladders together to provide longer sections unless they are specifically designed for such use.

- Keep ladders free of oil, grease, and other slipping hazards.

- Do not load ladders beyond the maximum intended load for which they were built, nor beyond their manufacturer's rated capacity.

- Use ladders only for the purpose for which they were designed.
- Keep the area around the top and bottom of ladders clear.
- Do not move, shift, or extend ladders while occupied.
- Inspect ladders for visible defects on a periodic basis and after any occurrence that could affect their safe use.
- Remove and store the ladder at the end of the workday.

5.2.0 SCAFFOLDS

Scaffolds are another piece of equipment often encountered by masons. Scaffolds are elevated working platforms that support workers and materials. One general safety rule for all types of scaffolds is that every scaffold should have a minimum safety factor ratio of 4:1. This means that the scaffolding should be able to support at least four times the weight that will be placed on it.

A typical scaffold is shown in *Figure 27*. The main part of the scaffold is the working platform. A working platform should have a guardrail system that includes a top rail, midrail, toeboard, and screening. To be safe and effective, the top rail should be approximately 42 inches high, the midrail should be located halfway between the toeboard and the top rail, and the toeboard should be a minimum of 4 inches high. If workers will be passing or working under the scaffold, the area between the top rail and the toeboard must be screened. Finally, the platform planks must be laid closely together. For safety purposes, the ends of the planks should overlap at least 6 inches and no more than 12 inches.

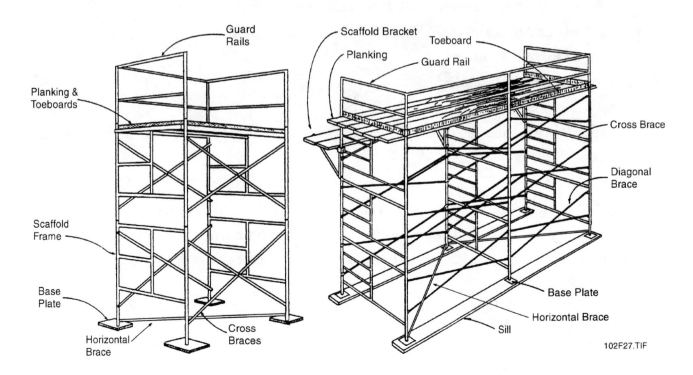

Figure 27. Tubular Steel Scaffold

Before using any scaffold, you must inspect it to make sure it is safe. Any defects or dangerous or unsafe conditions should be reported to your supervisor. If you are using a ladder to get on and off the scaffold, it is also important to follow all the ladder safety guidelines explained earlier in this module. The top of the ladder must extend at least 3 feet above the platform so that it will provide a hand grasp. Never climb on the cross braces of the scaffold. Braces are not made to support your weight.

When several people are working on a scaffold at the same time, you must make sure that everyone spaces themselves to distribute the weight evenly. And, when materials are delivered to the scaffold by forklift, keep everyone in the clear during the lifting. Be careful not to overload the scaffold with pallets of masonry materials at any one place.

The following additional guidelines should be followed in order to ensure safe scaffold use:

- Always raise the scaffold to the desired height. Do not try to increase height by using ladders, boxes, or sawhorses on top of a scaffold.
- Always keep scaffold planks clear of extra tools and materials.
- Always clean up any slippery substances that get spilled on a scaffold.
- Use extreme caution when working near power lines.
- Anchor freestanding scaffolds by guy wires in order to prevent tipping or sliding.
- Locate and read the posted safety rules and regulations for scaffolding use.
- When objects and tools are to be removed from the work area, lower them to the ground with a rope. Never throw or drop them from the scaffold.
- Remove all personnel from the platform before moving it.
- Keep tools and materials back from the edge of work platforms.
- Observe OSHA regulations.
- Use only approved scaffolding.
- Inspect scaffolds regularly.
- Install all braces and accessories according to the manufacturer's recommendations.
- Keep scaffolds plumb and level.
- Do not overload scaffolds.
- Be sure all safety rails are installed according to OSHA regulations.
- Be sure wheels are locked before ascending or disassembling rolling scaffolds.
- Always check for electrical wires and disengage wires before erecting metal scaffolds in the vicinity.
- Do not work on scaffolds during storms or high winds, or when scaffolds are covered with ice or snow.

- Construct and use only safe scaffolds. If a scaffold is wobbly or bouncy, or can be pulled down easily, it is not safe. A scaffold should have a sound structure and have toeboards and handrails. A scaffold's wooden floor planks must be free of knots, and broken members must be repaired immediately.

5.3.0 PERSONNEL LIFTS

Personnel lifts must meet all OSHA and ANSI requirements. For example, all hoist towers that are outside the structure should be enclosed for the full height on the side or sides used for entrance and exit to the structure. At the lowest landing, the enclosure on the sides not used for exit or entrance to the structure should be enclosed to a height of at least 10 feet. Other sides of the tower adjacent to floors or scaffold platforms should be enclosed to a height of 10 feet above the level of such floors or scaffolds. Towers that are inside structures should be enclosed on all four sides throughout the full height.

Additional requirements specify the type of anchoring to be used, the size and placement of hoistway doors and gates, and the enclosure and capacity of cars. Because it is human nature to ride rather than to climb stairs, be sure that any material hoist used on the job site is designed so that it will have the safety equipment needed to carry people. This will ensure a safe ride for all involved.

6.0.0 TOOL AND EQUIPMENT SAFETY

Many injuries result from improper or unsafe use of hand tools and small power tools. Becoming familiar with the basic tools and materials of your trade and understanding how to handle them safely are important steps toward working safely. The need to keep tools clean and in good working order cannot be overstressed. Safe and proper use of tools will not only help preserve the tools in good working condition but can also help to prevent accidents. All tools are capable of inflicting injury—even hand tools.

6.1.0 HAND TOOLS

The tools that masonry craftworkers must know how to use are many and skill in using them must be developed before they can work efficiently. Faulty, damaged, or improperly used tools and equipment pose a substantial threat to the safety and well-being of employees.

One of the most important rules regarding tool use is to use quality tools. Good-quality tools will last longer if they are used and maintained properly. Next, it is important to be sure to use the correct tool for the job. If you use the proper tool, the job will not require as much muscle power.

When using tools, you need to stay alert. This means positioning all cutting edges away from your body, keeping your fingers away from any and all cutting edges, and holding cutting tools in such a way that you will not be hurt if there is a slip. When working around electrical equipment, you should use tools with insulated or wooden handles. You also need to inspect tools frequently to make sure they are in good condition. Do not use a file, for example, unless it has a handle. And never use broken tools or tools with loose or broken handles or dull blades. If you find a broken or damaged tool, repair it promptly. Dispose of broken tools that cannot be repaired.

It is also important to maintain all tools in safe working order and keep them clean, dry, and sharp. The quality and quantity of work increases when using sharp tools in good condition. When tools are not cleaned before mortar has time to harden on them, the job of getting them back into readiness becomes more difficult and takes time away from the job at hand.

When not in use, it is important to store tools properly. Do not carry tools in your pockets and do not place tools where they can fall. After cleaning, tools that are prone to rust should be coated with a rust-preventing oil and stored in a dry place, with their cutting edges protected from damage. Any defective tools, machines, or other equipment should be tagged and reported to your supervisor.

6.2.0 POWER TOOLS

Power tools may be operated by electricity, air, or gasoline engines. Like hand tools, they should be operated in a safe manner at all times and stored properly when not in use. It is also very important to wear appropriate safety equipment and protective clothing when using power tools. Do not wear loose fitting clothes that can become caught in moving parts. You should also tie back long hair, roll up any long sleeves, and tuck in shirttails and pants legs.

If used improperly, power tools can cause fire, electric shock, and other injuries. It is very important, then, that you operate only those power tools that you have been trained to operate, and use power tools only for their intended use. You also need to be familiar with the correct operation and adjustments of power tools. And, do not attempt to operate a power tool that you have never used before or one for which you have not received instruction in its use.

Safe and proper operation of power tools calls for following these guidelines:

- Check to make sure an electric power tool is properly grounded before using it.
- Assume a safe and comfortable position before starting a power tool.
- Do not distract others or let anyone distract you while operating a power tool.
- Disconnect power before performing maintenance or changing accessories.
- Do not use dull or broken accessories. Check that saw blades are sharp and undamaged. Replace dull or broken blades.
- Do not leave power tools running and do something else.
- Report any unusual noises, sounds, or vibrations to the supervisor.
- Do not use a power tool that has had its guards or safety devices removed.

- Use a proper extension cord of sufficient size to service the particular electric tool you are using.

- Do not operate an electrical power tool if your hands or feet are wet.

- Check the insulation on cords and the condition of plugs and sockets every day. If cords, plugs, or sockets are frayed, worn, cut, or broken, repair them before using.

- String temporary extension cords and power lines so they will not create a tripping hazard and so they are protected from physical damage.

- Clean powered equipment with a brush. Do not clean equipment with your hands or with a rag. Do not clean equipment while it is running.

- Limit the exposure of connectors and tools to excessive moisture by using watertight or sealable connectors.

- Before using a drill, splitter, or grinder, be sure all electric wires, gas lines, and high-pressure lines are out of the way.

6.3.0 GASOLINE-POWERED TOOLS

These are additional safety guidelines that should be followed when working with gasoline-powered tools:

- Be sure there is proper ventilation before operating gasoline-powered equipment indoors.

- Use caution to prevent contact with hot manifolds and hoses.

- Be sure the equipment is out of gear before starting it.

- Always keep appropriate fire extinguishers near when filling, starting, and operating gasoline-powered equipment. OSHA requires that gasoline-powered equipment be turned off prior to filling with gasoline.

- Do not pour gasoline into the carburetor or cylinder head when starting the engine.

- Use the recommended starting fluid.

- Never pour gasoline into the fuel tank when the engine is hot or when the engine is running.

- Do not operate equipment that is leaking gasoline.

6.4.0 POWDER-ACTUATED TOOLS

Powder-actuated tools present additional safety concerns. These tools are designed for such things as driving fasteners into concrete or masonry. The powder-actuated tool, or nailing gun, gets its name from its source of power—a small charge of gunpowder. It is usually equipped for safety so that you have to press the end of the tool against the work surface before you can fire it. It fires just like any other gun—by pulling the trigger. The fastener is driven like a bullet through the workpiece and into the structure behind it.

This tool can use many different cartridges and powder loads, depending on the hardness of the structure into which the fasteners must be driven. The most important thing for you to remember about these powder-actuated tools is that they are extremely dangerous. Because of this, it is important to treat powder-actuated tools exactly as you would any other gun. Do not handle one unless you are being supervised by an authorized instructor. You must be trained and possess a valid operator's certificate before you can actually operate such a tool yourself.

Follow these safety guidelines before using powder-actuated tools:

- Stay out of the danger zone, which is greater for this tool than for any other. A misfired stub nail can fly hundreds of feet and still have enough force to seriously injure or even kill you.

- Dress properly. Wear your hard hat at all times. Also wear heavy boots and eye, ear, and face protection.

- Do not use powder-actuated tools for driving nails or spikes in walls, ceilings, or floors when people are working on the other side.

- Locate the tool's safety locking switch and be sure of the locked position.

- Determine whether the tool is loaded, either with a powder charge, a fastener, or both.

- Always treat the tool as though it were a loaded gun ready to fire.

- Always point the tool away from people, including yourself.

6.5.0   PRESSURE TOOLS

Pneumatic saws and grinders should have automatic overspeed controls so as to avoid runaway speeds that can cause carborundum saw discs and buffers to disintegrate. Guards should also be in good condition in order to restrain flying particles.

All tools that produce dust need exhaust systems to collect the dust in order to prevent contamination of the air in the work areas. Automatic emergency valves should be installed at all compressed air sources to shut off the air immediately if the hose becomes disconnected or is severed. This will help prevent wild thrashing of the hose.

SUMMARY

Safety is everyone's responsibility. You cannot assume that someone else is taking care of it. For a safe workplace, you and every worker on the job must be safety conscious and practice safe work habits. In order to work safely, you need to learn the safety gear and clothing required for each job, the safety rules and procedures associated with each job, general work site safety, the hazards associated with each job, and materials handling safety as it affects each job.

Safety gear and clothing, when worn appropriately, can help protect you and your fellow workers from harm. Typical safety gear includes a hard hat, safety goggles, ear protectors, and gloves. When working in areas of dust or toxic fumes, it will also be necessary to wear a respirator.

Of course, it is not enough to have the proper gear; you also need to act responsibly and think about what you are doing. To prevent accidents, for example, you should confine your hair in a ponytail or in your hard hat. You should also remove all jewelry and be sure to wear close-fitting clothing that will not get caught in machinery. When working in confined spaces, you should also be sure to use a carbon monoxide meter to warn you of dangerous levels of this deadly gas.

Each work site has its own hazards. That is why it is so important to follow general safety precautions such as keeping the area clean and free of obstructions, using safety harnesses where necessary, and familiarizing yourself with the location of firefighting equipment. You need to do everything possible to make the workplace as safe as it can be for yourself and others.

In order to prevent accidents, all work must be conducted in compliance with federal, state, and local laws; company policies and procedures; manufacturer's recommendations; and plain common sense. You must also follow OSHA guidelines for working in and around trenches and excavations, in controlled access zones, on scaffolding and ladders, and in confined spaces. OSHA has very strict requirements governing these areas. It is your job to know these requirements and to follow them.

Finally, you should familiarize yourself with the power tools and equipment you will be using on the work site. Read the operator's manual for each item and be sure to operate it only after you have received proper training. For safe and efficient operation, it is also important to maintain your tools and equipment in good working condition. This will ensure that they are always ready and safe to use you when you need them.

References

For advanced study of topics covered in this task module, the following reference materials are suggested:

Building Block Walls—A Basic Guide, National Concrete Masonry Association, Herndon, VA, 1988.

Masonry Skills, Third Edition, Richard T. Kreh, Delmar Publishers, Inc., Albany, NY, 1990.

Masonry Design and Detailing—For Architects, Engineers and Contractors, Fourth Edition, Christine Beall, McGraw-Hill Publishing, New York, NY, 1997.

The ABCs of Concrete Masonry Construction, Videotape 13:34 minutes, Portland Cement Association, Skokie, IL, 1980.

ACKNOWLEDGMENTS

Figures 1, 3, 12, 14, and 23-25 courtesy of Roy Jorgensen Associates, Inc.
Figure 2 courtesy of BEC Allwaste, Inc.
Figures 4-9, 13, 15, 16, 18-22, 26, and 27 courtesy of Associated General Contractors
Figures 10 and 11 courtesy of Miller Equipment Co.
Figure 17 courtesy of TSI, Inc.

REVIEW/PRACTICE QUESTIONS

1. Studies show that there is a much greater risk for injury when workers _____.

   a. become comfortable in performing skills
   b. follow prescribed safety procedures
   c. employ safe work habits
   d. attend company safety meetings

2. What is the leading cause of worker fatalities in the construction industry in the United States?

   a. Falls
   b. Accidents involving vehicles
   c. Objects falling on workers' heads
   d. Carbon monoxide poisoning

3. OSHA standards are enforced by _____.

   a. your employer
   b. your fellow employees
   c. federal and state safety inspectors
   d. insurance companies

4. A ladder must be provided at a construction site when _____.

   a. it is the only means of access from a working area for 25 or more employees
   b. it is the only means of exit from a working area for 25 or less employees
   c. a building or structure has two or more points of access between levels
   d. it is the only means of access from a working area for 25 or less employees

5. According to OSHA, an excavation _____.

   a. must be covered
   b. is a narrow hole made below the surface of the ground at least 4 feet deep
   c. is at least 15 feet wide
   d. is any man-made cut, cavity, trench, or depression in the earth's surface

6. Employees in an excavation are required to wear protective equipment when _____.

   a. the excavation is made entirely in stable rock
   b. the excavation is more than 5 feet deep
   c. a competent person has examined the ground and stated that there is no indication of a potential cave-in
   d. the excavation is less than 5 feet deep

7.   The color _____ identifies out-of-order signs on equipment.

   a.   blue
   b.   green
   c.   black
   d.   white

8.   Lanyards should be used to limit a fall to how many feet?

   a.   3
   b.   6
   c.   9
   d.   12

9.   Formal training programs _____.

   a.   are referred to as toolbox meetings
   b.   are always held during lunch
   c.   should be ignored
   d.   provide periodic refreshers on popular safety topics

10.   When working near dust or fumes, it is especially crucial for you to wear _____.

   a.   a respirator
   b.   safety goggles
   c.   hearing protectors
   d.   safety shoes

11.   Portland cement is hygroscopic, which means that _____.

   a.   it absorbs moisture
   b.   it is a powder
   c.   no moisture can affect it
   d.   anyone can use it to make mortar or concrete

12.   One of the easiest ways to promote safety on the work site is to _____.

   a.   attend OSHA meetings
   b.   practice practical jokes
   c.   keep the work site clean and free of obstructions
   d.   read all Material Safety Data Sheets

13.   Which of the following is not a good safety practice?

   a.   Assisting other trades when necessary to maintain safe operations
   b.   Storing materials and supplies in their proper places
   c.   Lifting heavy objects with your legs
   d.   Using compressed air to clean your clothes

14. Which of the following things should you do before entering a newly dug trench or excavation?

    a.   Check nearby structures to see how they will affect the excavation

    b.   Provide ladders every 10 feet when the trench is four feet deep or more

    c.   Wear Appropriate Personal Protective Equipment

    d.   Move all heavy trucks and equipment to the edges of the excavation

15. A controlled access zone _____.

    a.   allows workers to work without the use of conventional fall protection systems

    b.   requires all workers within it to wear conventional fall protection systems

    c.   permits free access to all workers

    d.   restricts access to only certain supervisors

16. What type of fire extinguisher is designed to be used to control fires that occur with iron, magnesium, sodium, and potassium?

    a.   Class A

    b.   Class B

    c.   Class C

    d.   Class D

17. GFCIs _____.

    a.   protect workers from line-to-line contact hazards

    b.   protect workers from the possibility of a ground fault

    c.   increase electrical hazards

    d.   increase circuit tripping

18. Which of the following types of information for hazardous materials is not provided by the MSDS?

    a.   Specific health hazards

    b.   Steps for emergency first aid

    c.   Company-specific procedures for handling the material

    d.   Physical and chemical properties data

19. Toxic substances can cause terminal diseases such as _____.

    a.   shock

    b.   headache

    c.   coughing

    d.   emphysema

20. Approximately _____ percent of occupational injuries occur when handling or moving construction materials.

    a. 25
    b. 50
    c. 75
    d. 100

21. Some _____ of construction injuries and deaths involve falls from elevations.

    a. one-fourth
    b. one-third
    c. one-half
    d. two-thirds

22. Which of the following is a correct safety procedure for using a ladder?

    a. Place the base of the ladder on 4 × 4 blocks.
    b. Keep the rungs painted.
    c. Never overreach from a ladder.
    d. Always place the ladder no more than 4 feet from the structure.

23. How much clearance should be around a ladder?

    a. 30 inches in back and 6 inches in front
    b. 2 feet in back and 1 foot in front
    c. 6 inches in back and 30 inches in front
    d. 2 feet in front and 1 foot in back

24. Every scaffold should have a minimum safety factor of _____.

    a. 1:1
    b. 2:1
    c. 3:1
    d. 4:1

25. When using powder-actuated tools, it is important to _____.

    a. locate the tool's safety locking switch and be sure of the locked position
    b. remove your hard hat
    c. use one hand
    d. always keep them loaded

ANSWERS TO REVIEW/PRACTICE QUESTIONS

<u>Answer</u>		<u>Section Reference</u>
1.	a	1.1.0
2.	a	2.0.0
3.	c	2.2.0
4.	a	2.2.2
5.	d	2.2.3
6.	b	2.2.3
7.	a	2.2.6/Figure 1
8.	a	2.2.7
9.	d	2.3.0
10.	a	2.4.1
11.	a	3.8.0
12.	c	3.1.0
13.	d	3.1.0
14.	c	3.2.0
15.	a	3.4.0
16.	d	3.6.0
17.	b	3.7.0
18.	c	4.0.0
19.	d	4.1.0
20.	a	4.3.0
21.	b	5.0.0
22.	c	5.1.0
23.	c	5.1.0
24.	d	5.2.0
25.	a	6.4.0

NCCER CRAFT TRAINING USER UPDATES

The NCCER makes every effort to keep these manuals up-to-date and free of technical errors. We appreciate your help in this process. If you have an idea for improving this manual, or if you find an error, a typographical mistake, or an inaccuracy in the NCCER's Craft Training Manuals, please write us, using this form or a photocopy. Be sure to include the exact module number, page number, a description of the problem, and the correction, if possible. Your input will be brought to the attention of the Technical Review Committee. Thank you for your assistance.

Instructors – If you found that additional materials were necessary in order to teach this module effectively, please let us know so that we may include them in the Equipment/Materials list in the Instructor's Guide.

Write: Curriculum and Revision Department
National Center for Construction Education and Research
P.O. Box 141104
Gainesville, FL 32614-1104

Fax: 352-334-0932

Craft _____ Module Name _____

Module Number _____ Page Number(s) _____

Description of Problem

_____

_____

_____

_____

(Optional) Correction of Problem

_____

_____

_____

(Optional) Your Name and Address

_____

_____

_____

Tools and Equipment
Module 28103

NATIONAL
CENTER FOR
CONSTRUCTION
EDUCATION AND
RESEARCH

TOOLS AND EQUIPMENT

OBJECTIVES

After completing this module, the trainee will be able to:

1. Identify and name the tools used in performing masonry work.
2. Identify and name the equipment used in performing masonry work.
3. Describe how each tool is used.
4. Describe how the equipment is used.
5. Associate trade terms with the appropriate tools and equipment.
6. Demonstrate the correct procedures for assembling and disassembling scaffolding according to federal safety regulations, under the supervision of a competent person.

Prerequisites

Successful completion of the following Task Modules is recommended before beginning study of this Task Module: Core Curricula; Masonry Level 1, Modules 28101 and 28102.

Required Trainee Materials

1. Trainee Task Module
2. Pencil and paper
3. Appropriate Personal Protective Equipment

COURSE MAP

This course map shows all of the task modules in the first level of the Masonry curricula. The suggested training order begins at the bottom and proceeds up. Skill levels increase as a trainee advances on the course map. The training order may be adjusted by the local Training Program Sponsor.

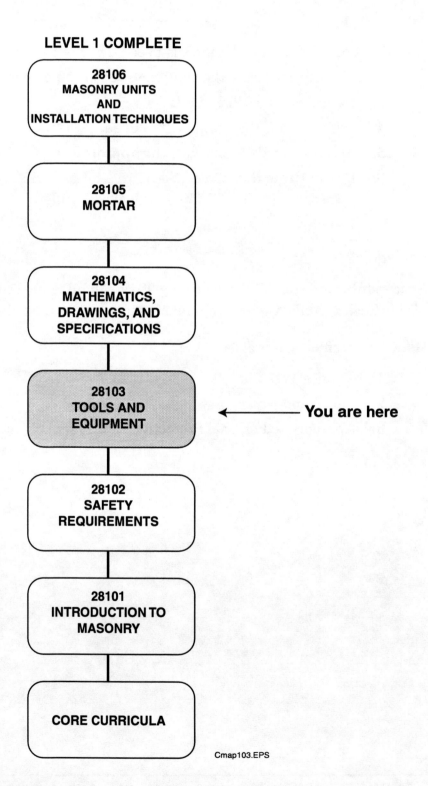

LEVEL 1 COMPLETE

28106
MASONRY UNITS
AND
INSTALLATION TECHNIQUES

28105
MORTAR

28104
MATHEMATICS,
DRAWINGS, AND
SPECIFICATIONS

28103
TOOLS AND
EQUIPMENT ← You are here

28102
SAFETY
REQUIREMENTS

28101
INTRODUCTION TO
MASONRY

CORE CURRICULA

Cmap103.EPS

TABLE OF CONTENTS

TABLE OF CONTENTS (Continued)

Trade Terms Introduced In This Module

Bed joint: A horizontal joint between two masonry units.

Buttering: Applying mortar to the ends of a brick or block masonry unit.

Course: A row or horizontal layer of masonry units.

Cube: A strapped bundle of approximately 500 standard bricks, or 90 standard blocks, usually palletized. The number of units in a cube will vary according to the manufacturer.

Footing: The base for a masonry unit wall, or for a concrete foundation, that distributes the weight of the structural member resting on it.

Grout: A compound of portland cement, lime, fine aggregates, and water, with a high enough water content that it can be poured into spaces between masonry units.

Head joint: A vertical joint between two masonry units.

Lead: The two corners of a structural unit or wall, built first, and used as a position marker and measuring guide for the entire wall.

Masonry unit: A building block made of brick, cement, stone, dressed stone, clay, adobe, rubble, glass, or any other material, that is bonded with mortar into a structural unit.

Parge: A thin coat of mortar or grout on the outside surface of a wall. Applying such a thin coat or mortar. Parging prepares a masonry surface for attaching veneer or tile, or parging can waterproof the back of a masonry wall.

Pointing: Troweling mortar or a mortar-repairing material, such as epoxy, into a joint after masonry is laid.

Retempering: Adding water to mortar to replace evaporated moisture and restore proper consistency. Any retempering must be done within the first 2 hours after mixing, as mortar begins to harden after 2½ hours.

Skewback: A sloping surface against which the end of an arch rests; may be brick cut on an angle.

Temper: To remix mortar by adding water to make it more workable.

Trestle: A system of scaffolding with diagonal legs; a split-leg support for a system of scaffolding.

Tuckpointing: Filling fresh mortar into cutout or defective joints in masonry.

Why are tools so important for masonry work? Brick cannot be cut with bare hands, and mortar cannot be laid with fingers. Masonry tools are the interface between the worker and the work. In fact, the quality of the tool—especially the trowel—affects the quality of the work. Because of the many things that masons do, there are many special-purpose tools. Some hand tools are unchanged since ancient times, while some power tools use the most up-to-date technology.

This module introduces the tools and equipment that you will need to lay **masonry units**. Specifically, this module addresses hand tools, measuring tools, hand-powered equipment, power tools, and power equipment. By the end of this module, you will be able to identify each item and what it does. It includes common as well as some uncommon tools and equipment. Remember that not every tool will be used on every job.

1.1.0   TOOL SAFETY

Tool and equipment safety falls into four categories:

* Following safe work practices and procedures
* Inspecting tools and equipment before use
* Using tools and equipment properly
* Keeping tools and equipment clean and properly maintained

Safety for tools requires using them properly, keeping them clean, and being careful not to damage them. Damaged tools or equipment can break and cause injuries as well as slow down or stop work. To avoid damaging tools, remember to use the right tool for the job. Some examples of the misuse of tools are listed below:

* A trowel is not a hammer, so do not hammer with it.
* Do not hammer on your plumb rule.
* Do not use your plumb rule as a pry bar.
* Do not use broken or defective tools.
* Do not cut reinforcement with a hammer, chisel, or trowel.

CAUTION: Wear safety goggles when cutting or chipping any type of masonry unit.

Accidents also happen when you leave tools and equipment in the way of other workers. Store tools safely, where other people cannot trip over them and the tools cannot get damaged. Clean, well-kept tools make for safe work as well as good work.

Do not drop or temporarily store tools or masonry in pathways or around other workers. Stack masonry units neatly to avoid the risk of them toppling over. Stack materials by reversing direction on every other layer, so they will be less prone to tip. Keep the pile neat and vertical to avoid snagging clothes. Do not stack masonry units higher than your chest (as a rule). A stack that is too high is more likely to tip over.

CAUTION: Mortar and **grout** can be very caustic. It is important to protect your skin and eyes when working with mortar and grout. Wear gloves and other Appropriate Personal Protective Equipment as required for the specific job.

1.2.0 TOOL MAINTENANCE

Tools and equipment must be in good repair to be safe. Repair or replace defective tools and equipment immediately.

Clean all tools and equipment that touch mortar or grout immediately after use. Use a bucket of water to soak tools that you will use again in a few minutes (keep wooden handles out of the soaking water). If mortar is left to dry on a tool it will harden and make the tool unusable.

Wash tools thoroughly with water and a wire brush. Be sure that you remove all mortar or grout completely. If you do not use tools regularly, coat them lightly with oil, grease, or other approved coating to prevent rust and corrosion. After cleaning, check that tool handles are secure and free from cracks or splinters. Oil the pivot joints and wooden parts of the tools. Sharpen blades and cutting edges when they become dull or nicked.

Clean all motorized tools and equipment after use. Always turn off and disconnect the power before cleaning or fixing powered tools and equipment. Check with your supervisor when in doubt about the care of specific items.

Here are some specific hand tool care tips:

- Keep your mason's hammer sharp and square
- Keep folding rules oiled
- Keep chisel blades sharp and chisel heads and blades free of burrs
- Keep all tool handles tight

2.0.0 HAND TOOLS

The quality and use of masonry hand tools greatly affect the quality of the final work. Hand tools are used to coat, cut, carry, clean, align, and level masonry units. The following sections describe masonry hand tools you will typically use to perform masonry construction.

2.1.0 TROWELS

The mason's trowel comes in many sizes and shapes. The basic trowel, *Figure 1*, consists of a steel blade ground to the proper balance, taper, and shape. The narrow end of the blade is the point, and the wide end is the heel. The blade is connected to the handle by a shank. The handle is made of wood or plastic. The wood handle has a band, or ferrule, on its shank end to prevent splitting. When holding a trowel, keep your thumb along the top of the handle, not the shank. If your thumb is on the shank, it can get coated with mortar as you work.

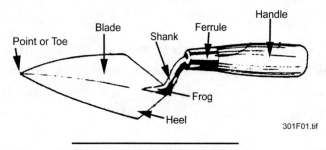

301F01.tif

Figure 1. Parts Of A Trowel

The blade can come with a sharply angled heel (London pattern) or a square heel (Philadelphia pattern), as shown in *Figure 2*. Trowels come in different shapes and sizes for different purposes.

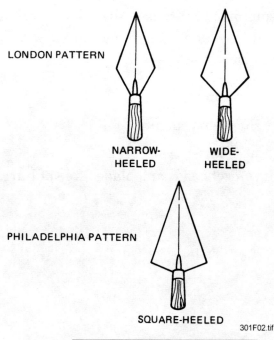

301F02.tif

Figure 2. Brick Trowel Shapes

Trowels can range in width from about 4" to 7", and can be up to 13" long. *Figure 3* shows different types of trowels. These trowels have specialized uses and will not be used as often as a standard brick trowel.

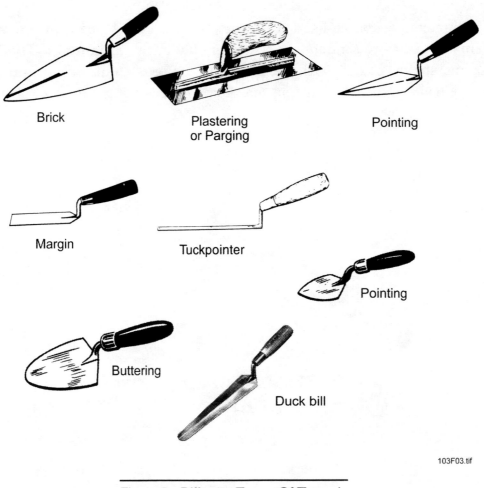

Figure 3. Different Types Of Trowels

The standard or brick trowel is the all-purpose model, used for cutting, mortaring, and placing units. The **buttering** trowel is extra-wide for applying mortar. The small **pointing** trowel fits into tight spaces. It points, grouts, and tools mortar joints. The margin trowel smoothes joints and mixes grout or other compounds. The plasterer's or parging trowel applies a thin coat, or **parge**, of mortar. The parge coat will hold tile or veneer to the masonry units. The **tuckpointing** trowel comes in widths ranging from ³⁄₁₆" to 1", to fit commonly-used joint sizes.

2.2.0 HAMMERS

Mason's hammers fall into two categories: cutting hammers and mauls. Using the right hammer for each job will ensure the most efficient use of your energy. Many types of hammers come with wooden handles. Inspect these handles every day to make sure they are not loose, splintered, or cracked. Chisel edges on hammers need to be inspected and sharpened just as chisels do.

CAUTION: If the wooden handle on a hammer becomes loose, replace it immediately.

2.2.1   Cutting Hammers

The brick hammer is a double-headed hammer with a chisel head on one side, shown in *Figure 4*. This is the most commonly used, everyday mason's hammer. The brick hammer drives nails and strikes chisels with one head and can break, cut, or chip masonry material with the other head. Use it for cutting masonry units, setting line pins, and nailing wall ties.

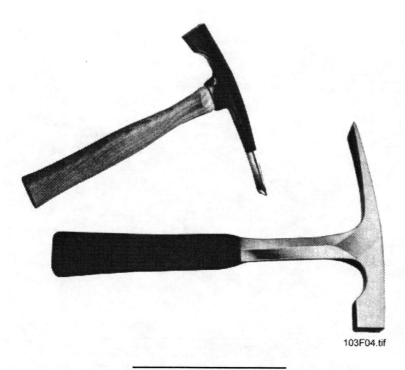

103F04.tif

Figure 4. Brick Hammers

Brick hammers are one-piece, drop-forged steel tools, or steel heads on wooden shafts. Steel hammers need a comfortable grip sleeve over the shaft. The weight of this hammer usually ranges from 12 to 24 ounces. If you are buying a hammer, pick a weight that is comfortable for your hand.

The tile hammer is a smaller brick hammer with a smaller, thinner cutting head. *Figure 5* shows this hammer, which usually weighs about 9 ounces. Use it for cutting and trimming tile where you need more precision than you can get with a brick hammer. Do not drive nails, strike chisels, or do other heavy work with a tile hammer.

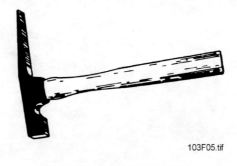

Figure 5. Tile Hammer

The stonemason's hammer, shown in *Figure 6*, resembles an ax. This heavy hammer has a cutting blade on one end of the head. Use this special-purpose hammer for dressing, cutting, splitting, or trimming stone.

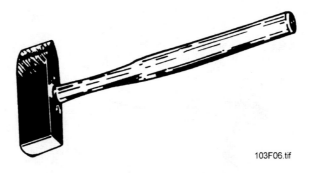

Figure 6. Stonemason's Hammer

The cutting end on all chisel-headed hammers should be sharpened by a blacksmith. Grinding it on a wheel will affect the temper of the head. If the head loses its temper, it can shatter without warning.

CAUTION: Never strike the heads of two hammers together.

2.2.2   Mashes And Mauls

Mashes and mauls are heavy club-headed hammers used for striking. They usually have two striking heads.

The mason's mash is a drop-forged, one-piece tool with a grip sleeve over the shaft. Or, it can have a steel head mounted on a wooden handle. As shown in *Figure 7*, it resembles a short-handled, double-headed sledgehammer. The mash is most commonly used to strike chisels or to cut masonry units. This hammer is too heavy to use for setting line pins or nailing wall ties.

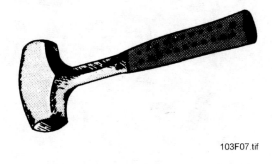

Figure 7. Mash

The bushhammer, or maul, is a square-ended, rectangular, double-headed sledgehammer. It has a heavy head and a long wooden handle, as shown in *Figure 8*. It is a stonemason's tool with toothed ends, used to face block, stone, or concrete.

Figure 8. Bushhammer

The rubber mallet has a rubber head with two flat ends and a wooden handle. This hammer, shown in *Figure 9*, can tap or drive something without leaving marks. Use the rubber mallet for setting stone, marble, tile, or other finely finished units into place.

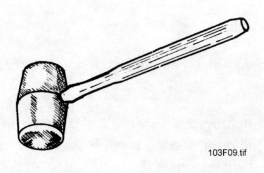

Figure 9. Rubber Mallet

When working with wooden-handled hammers, be sure to inspect them daily. Check that the handle is securely seated in the head. If the handle is loose, replace it before using the hammer.

2.3.0 CHISELS

Masons use chisels to cut masonry units. For everyday work, masons use a hammer to cut brick or block, but for precise, sharp edges, they use a chisel. There are several types of chisels, as shown in the following figures.

Chisels are cutting tools, so keep them sharp and tempered. The best way to keep chisels sharp is to take them to a blacksmith.

Chisels become damaged after prolonged use. Cutting edges get notches or burrs from striking rough surfaces or metal. Heads flatten after long use. The striking head mushrooms out, and metal burrs form at the edges of the head; these deformed edges can fly off and may cause injury. Inspect your chisels every day for dullness or deformation.

CAUTION: Do not use a chisel with a deformed blade or a mushroomed head. Repair it or replace it immediately.

To prevent injury, grind off the deformed part of the chisel head on a grinding wheel. As you grind, cool the chisel with water to keep it from overheating. If it gets overheated, it will lose the temper of its steel and unexpectedly shatter. Do not repair or sharpen the blade yourself; take it to a blacksmith.

Figure 10 shows a mason's steel chisel, used for general cutting. It has a narrow blade for a neat, clean cut. Use it to cut out masonry and make repairs. Use it for brick, block, and veined stone.

103F10.tif

Figure 10. Mason's Chisel

Figure 11 shows a brick set. This is a wider chisel, up to 5" wide, with a thicker, beveled cutting blade. Traditionally, the brick set is as wide as the bricks it cuts. A wider version of the brick set is called the blocking chisel, or bolster chisel. This is usually 8" wide, and used for cutting block.

103F11.TIF

Figure 11. Brick Set And Blocking Chisel

The rubber-grip mason's chisel is shown in *Figure 12*. This chisel has a rubber cushion on the handle to soften the impact of the hammer blow. This chisel has a wider cutting blade than the standard mason's chisel and is used for cutting brick, block, or stone.

103F12.tif

Figure 12. Rubber-Grip Mason's Chisel

The tooth chisel, in *Figure 13*, has a toothed edge. This chisel is designed to cut soft stone and shape it to fit. The pitching tool, also in *Figure 13*, is for sizing, trimming, and facing hard stone.

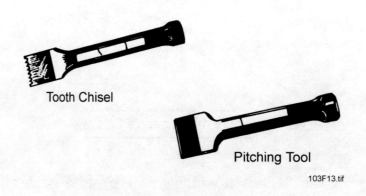

Tooth Chisel

Pitching Tool

103F13.tif

Figure 13. Tooth Chisel And Pitching Tool

The plugging or joint chisel, shown in *Figure 14*, has a sharply tapered blade. It is also called a tuckpointer's chisel. Use this chisel for cleaning mortar joints, for chipping mortar out of a joint, or for removing a brick or block from a wall.

103F14.tif

Figure 14. Plugging Or Joint Chisel

The difference between these chisels may not seem significant, but the differences will become important as you learn to use these tools.

2.4.0 JOINTERS

Jointers are steel tools for finishing or pointing the surface of mortar joints. Also called joint tools or finishing tools, the standard size jointer will fit into the standard joint between bricks or blocks. Jointers are available to fit a range of joint sizes, and come in various shapes to give different effects to the finished joint. *Figure 15* shows how the final surface appearance is determined by the shape of the jointer that is used.

Jointing compresses the mortar and decreases moisture absorption at the surface, so it adds water protection. Struck and raked joints are not recommended for exterior walls as they do not offer good water protection.

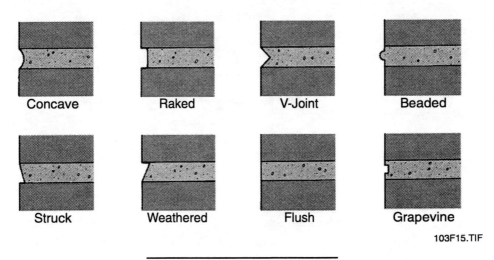

Concave   Raked   V-Joint   Beaded

Struck   Weathered   Flush   Grapevine

103F15.TIF

Figure 15. Tooled Mortar Joints

The jointers are usually cast or forged shaped steel rods, with or without wooden handles. *Figure 16* shows some simple jointers for compressing and waterproofing joints. These convex, round, flat, and v-jointers are used on the short vertical or head joints.

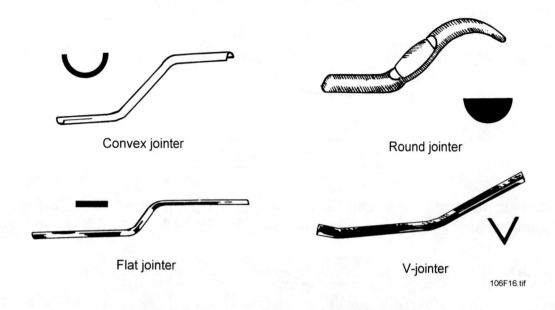

Convex jointer

Round jointer

Flat jointer

V-jointer

106F16.tif

Figure 16. Convex, Round, Flat, And V-Jointers

The flat jointer is sometimes called a slicker. Either end of a slicker can be used to shape mortar; usually the ends are different sizes. The slicker is handy for working on inside corners or other tight places where a neat joint is needed.

The **bed joints**, or the long horizontal joints, are tooled after the **head joints**. Longer jointers with wooden handles are used for the horizontal joints. As shown in *Figure 17*, these sled-runner jointers come in a variety of shapes also, to match the head joints.

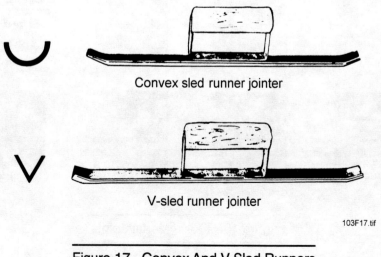

Convex sled runner jointer

V-sled runner jointer

103F17.tif

Figure 17. Convex And V-Sled Runners

Another type of jointer is the raker or rake-out jointer; these also come mounted on skate wheels. *Figure 18* shows skatewheel and regular rakers. The skatewheel raker is used for bed joints. The regular rake-out jointer is used for joints where the skatewheel raker will not reach. Note that the raking is done by an adjustable set screw, so different depths can be raked. The advantage of the skatewheel raker is the speed with which it forms a neat, hole-free joint.

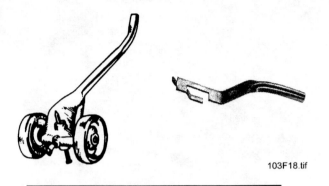

103F18.tif

Figure 18. Skatewheel And Hand Rakers

All jointers need to be cleaned after they are used. If mortar hardens on them, they are not usable for smoothing.

2.5.0 BRUSHES

Brushing masonry work removes any burrs or excessive mortar. This is the finishing process for the wall or floor. The brush should have stiff plastic or wire bristles. A stove brush, as shown in *Figure 19*, has a longer handle to keep fingers clear. Some masons use a floor brush sawed in half. Brushes are also useful for brushing off **footings** before laying masonry units and for cleaning the work area.

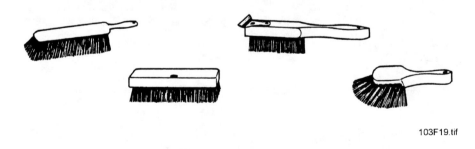

103F19.tif

Figure 19. Mason's And Stove Brushes

For cleaning stains, brushes are used to apply acid to the face of masonry units. A cleaning brush may also have a scraper on one edge (*Figure 19*). Acid brushes are made of stiff plastic or fiber and have longer handles. When applying acid, use a long-handled brush and wear gloves and eye protection.

2.6.0   BRICK TONGS

Brick tongs are designed to carry bricks without chipping or breaking them. As shown in *Figure 20*, they are made of an adjustable steel clamp with a locking nut and a handle. They can be adjusted for different sizes, and will hold 6 to 11 masonry units. With practice, the mason can carry loaded brick tongs in one hand, freeing the other hand for other activities.

103F20.tif

Figure 20. Brick Tongs

2.7.0   OTHER HAND TOOLS

Masons use pinch bars, caulking guns, bolt cutters, and grout bags in addition to their other tools. As shown in *Figure 21*, these tools will fit into the mason's tool bag. The pinch bar is used to pry out masonry units. The bolt cutter is used to cut reinforcement or ties. The caulking gun is used to add caulking to expansion joints. The grout bag, with a metal or plastic tip, is squeezed to apply grout between masonry units.

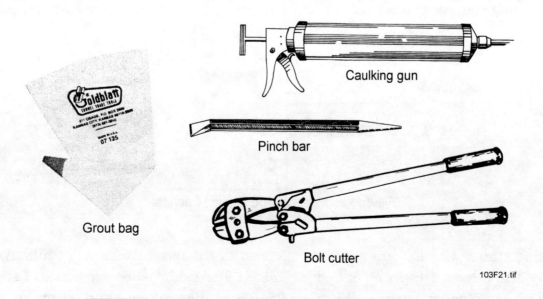

Caulking gun

Pinch bar

Grout bag

Bolt cutter

103F21.tif

Figure 21. Pinch Bar, Caulking Gun, Bolt Cutter, And Grout Bag

2.8.0 TOOL BAG

Figure 22 shows a mason's canvas tool bag with leather straps and handles. The tool bag keeps tools together and within reach as the mason moves about. Typically, a tool bag measures 14" to 18" across and has inside pockets for small items. The steel square and long level will fit under the leather straps on the outside of the bag. Some tool bag handles convert to shoulder straps so masons will have hands free for climbing ladders or scaffolding.

103F22.tif

Figure 22. Mason's Tool Bag

3.0.0 MEASURES AND MEASURING TOOLS

Measures and measuring tools are key to good masonry. They are used on every **course** to check that a wall is going up level and true. If they are not used constantly and consistently, it is easy to end up with a wall that is full of bulges and hollows, or "out of wind."

Because measures and measuring tools are so important to the final product, take time to learn how to read them accurately and quickly. Keep them clean, free of dust or grit. Buy the sturdiest, easiest-to-read, most accurate measures and measuring tools you can find.

Plumb rules are the most important tools in checking course alignment. Other measuring and aligning tools are rules, squares, lines, corner poles, chalk lines, and plumb bobs. These course alignment tools are discussed in the following sections.

3.1.0 LEVELS OR PLUMB RULES

Levels were introduced in the Core Curricula, *Introduction to Hand Tools*. The level, or plumb rule, establishes two measures:

- A plumb line that is vertical to the surface of the earth
- A level line that is horizontal to the surface of the earth

The plumb rule works by air bubbles in sealed vials filled with oil or alcohol, as shown in *Figure 23*. The bubbles come to center in the tubes within the marks to mark the level or plumb point.

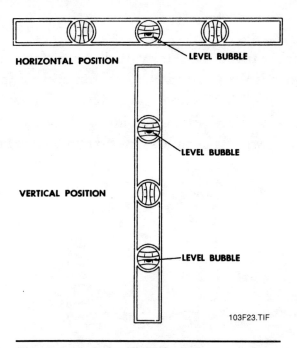

HORIZONTAL POSITION

LEVEL BUBBLE

VERTICAL POSITION

LEVEL BUBBLE

LEVEL BUBBLE

103F23.TIF

Figure 23. Spirit Bubbles In A Plumb Rule

Plumb rules can come with single or double sets of bubble vials. They are made of hardwood or metal and plastic as light as possible without sacrificing strength. The standard size runs 42" to 48" long. A smaller 18" to 24" size is used where the larger size will not fit. Masons use plumb rules continually as they lay masonry, to check the plumb and level of individual masonry units.

The plumb rule is usually the most expensive mason's tool. Always check the action of the bubble vials against another plumb rule before buying a plumb rule. Check the accuracy of your plumb rule daily, especially if it is dropped or jarred. Do not tap a plumb rule with a hammer to level a brick.

Clean your plumb rule carefully at the end of the day to keep any mortar from hardening on it. Wipe off wooden plumb rules with a rag dampened with linseed oil to preserve the wood. Clean a metal plumb rule with a dry cloth so dust and grit do not stick to it. You may want to oil it occasionally as well.

3.2.0 RULES

Masons use two kinds of rules, a 6-foot folding rule and a 10-foot retractable tape. Mason's rules have special marking scales on them, as shown in *Figure 24*.

The brick spacing rule has markings for different sizes of bricks. It shows the course spacing for each brick size. This measure is also called the course counter rule. Use it to lay out and space standard brick courses to dimensions that are not modular. It is useful for spacing mortar joints around door tops, window tops, or other places where differences must be figured.

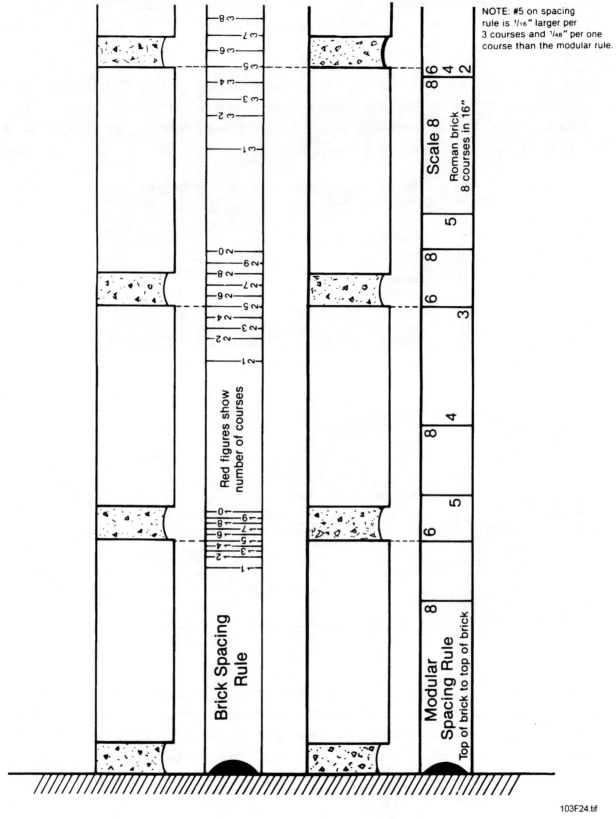

NOTE: #5 on spacing rule is ¹/₁₆" larger per 3 courses and ¹/₄₈" per one course than the modular rule.

Scale 8

Roman brick 8 courses in 16"

Red figures show number of courses

Brick Spacing Rule

Modular Spacing Rule

Top of brick to top of brick

Figure 24. Spacing Rules

103F24.tif

The modular spacing rule is based on a module of 4 inches. It has 6 scales ranging from 8 to 2, with the scale 6 representing the size of a standard brick. The modular rule can be used for block as well as brick.

The mason's steel tape and folding rule are both marked with spacing measures. They are available with either the modular or the course markings. *Figure 25* shows both these rules, which carry inch measurements as well. Folding rules are usually made of wood, with brass joints and pivots between the sections.

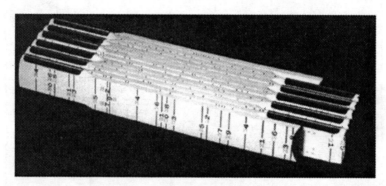

FOLDING RULE

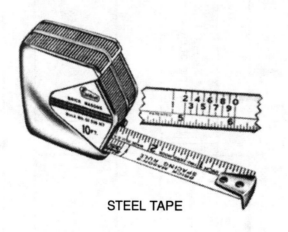

STEEL TAPE

103F25.TIF

Figure 25. Folding Rule And Steel Tape

Masons prefer a folding rule because one person can easily and accurately extend it to the needed length. Tapes must be stretched without bowing, which may need an extra hand over longer distances. Both rules and tapes should be inspected and cleaned daily. Use a very light touch of linseed oil on wood rules, and be sure to clean dust and grit from the pivots.

3.3.0 SQUARES

Masons use several types of squares. The framing or steel square, shown in *Figure 26*, looks like a carpenter's square. Use this for laying out **leads** and other corners, and for checking that corners are square.

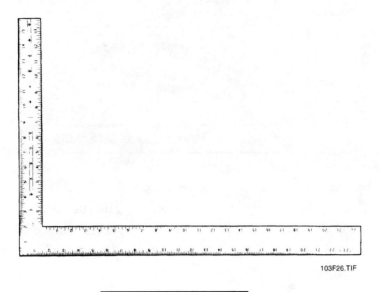

Figure 26. Steel Square

The combination square, or T-square, *Figure 27*, has a movable crosspiece and a built-in 45-degree angle. Use this for marking right-angle and 45-degree angled cuts, and for checking these angles on cuts. The bevel, to the right in *Figure 27*, has a set screw in the crosspiece. The screw fixes the angle of the crosspiece so that the same angle can be used for marking many cuts. The bevel is useful when the job calls for an odd angle other than 90 degrees or 45 degrees. Use the bevel for marking these kinds of cuts, for marking **skewbacks**, and for checking the angles on cuts.

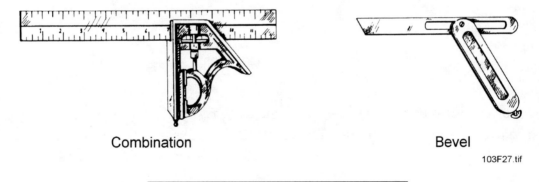

Combination

Bevel

Figure 27. Combination Square And Bevel

3.4.0 MASON'S LINE AND FASTENERS

The plumb rule is used to lay out structural masonry elements 4 feet in length or shorter. The mason's line is used to lay out masonry structures longer than 4 feet. The line itself is nylon cord, twisted or braided. The typical line is about 150 pounds test and should be used at 20 percent of its test. This means it is good for a pull of up to 30 pounds. Braided line is preferred because it will not sag as much when pulled tight and will last longer. Usually, the line is bought wound around a core. If it is rewound on a shuttle or holder as shown in *Figure 28*, it will not tangle as it is used and re-used.

103F28.tif

Figure 28. Mason's Line On Shuttle

The line is stretched tight as it is strung between two fixed points. For most jobs, the fixed points are on the already laid wall corners, or leads. The line becomes the guide for laying the course of masonry units between the leads, or fixed points. Using the line properly will result in a wall without bulges or hollows. The masonry units are laid under the line and the line is moved up for each course.

There are three methods of fastening the line in place: pins, blocks, and stretchers. Line pins (*Figure 29*) are about 4" long and made of steel. Drive them into the wall or structure at the marked point and string the line tightly between them. Line pins leave holes, and the string must be remeasured or retightened for each course. However, line pins are harder to accidentally knock off than line blocks.

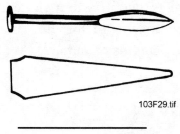

103F29.tif

Figure 29. Line Pins

Line blocks, or corner blocks, are shown in *Figure 30*, right. They are made of wood or plastic, with wood gripping the corner better. The knotted line passes through the slit in the back of the block. The line tension between the two blocks holds the blocks in place. Line stretchers are shown in *Figure 30*, left. They have a flatter profile and also use the line tension to hold them in place. They come in standard and adjustable sizes and fit more snugly to the lead than corner blocks.

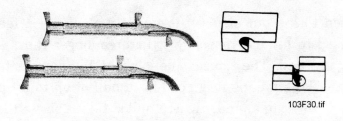

103F30.tif

Figure 30. Line Stretchers And Corner Blocks

Unlike pins, blocks and stretchers must have a corner to be snugged against. They do not leave holes in bricks or walls, but they do project slightly from the corners of the wall. They can be accidentally knocked off. Because of the tension on the line, a flying block or stretcher can severely injury someone in its path.

CAUTION: Use extra care working with corner blocks or stretchers to avoid knocking them off.

Once the line is in place, it may need more support. Line trigs, sometimes called twigs, are steel fasteners (*Figure 31*). They hold the line in position and keep it from sagging. A very long line may need several trigs, even if it is stretched tight. Masonry materials suppliers usually give away trigs.

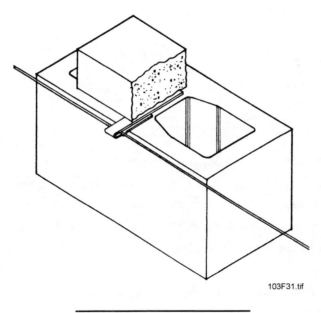

103F31.tif

Figure 31.   Line Trig Or Twig

The trig slips over the line and rests on a masonry unit that has been put in place to support it. A half brick or piece of block can anchor the trig to ensure the line is not disturbed.

3.5.0   CORNER POLE

The next in the series of course alignment measures is the corner pole, or deadman. As shown in *Figure 32*, the corner pole is any type of post braced into a plumb position so that a line can be fastened to it. This allows a wall to be built without building the corners first. This tool is useful when building a veneer wall. The corner pole can be braced against an existing part of a structure or the block foundation.

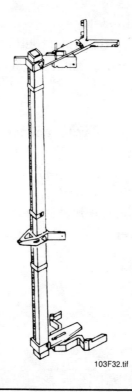

103F32.tif

Figure 32. Corner Pole Or Deadman

Corner poles can be made from dimensional lumber and braces. Commercial corner poles are metal with proprietary line blocks that can be attached to them. Commercial corner poles have masonry units marked on them. A hand-made corner pole can be marked off in masonry units with a grease pencil, marker, or carpenter's pencil. A corner pole with markings for different stories of a building is called a story pole.

3.6.0   CHALK LINE AND CHALK BOX

A chalk box (*Figure 33*) is a metal or plastic case with finely ground chalk and about 50 feet of twisted cotton line wound on a spool inside. As the line unreels through the opening in the box it picks up powdered chalk. The line can be stretched between two points and snapped. This will leave a chalk mark exactly under the line. Masons use chalk lines to establish straight lines for the first course of bricks in a wall because they can snap the chalk line against the foundation slab or footing.

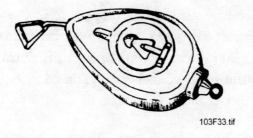

103F33.tif

Figure 33. Chalk Box

Before using the line every day, check that there is enough chalk in the reservoir and that the string is not frayed. Both of these parts should be replaced or refreshed periodically.

3.7.0 PLUMB BOB

Sometimes plumb is marked with a plumb bob, shown in *Figure 34*. The plumb bob is a pointed weight at the end of a length of mason's line. The length of the line is easily changed. A plumb bob can establish vertical plumb points. Use it to mark a point directly under another measured point. A plumb bob is also useful for checking the plumb of a story pole or corner pole.

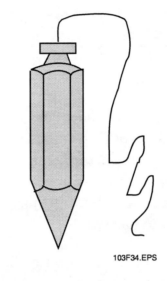

103F34.EPS

Figure 34. Plumb Bob

4.0.0 HAND-POWERED MORTAR EQUIPMENT

Hand-powered mortar equipment is used on small and large jobs. These pieces of equipment complement the hand tools in that they complete the tool set needed to build a structural masonry unit. The following sections introduce this equipment.

4.1.0 MORTAR BOXES

Mortar is mixed in a steel or plastic mortar box (*Figure 35*). The average mortar box is about 32" × 60" and sits on the ground. Make sure the box is set level so that water does not collect in one end. After it is mixed, the mortar may be moved to a smaller mortar board or mortar pan, also shown in *Figure 35*.

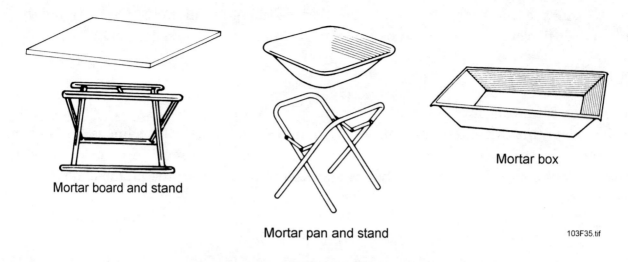

Mortar board and stand

Mortar pan and stand

Mortar box

103F35.tif

Figure 35. Mortar Box, Board, And Pan

The mortar pan and mortar board can be located next to the working mason. They sit on stands so they are convenient for working higher courses. The metal or plastic mortar pan fits more securely on the stand, but mortar tends to get stuck in its corners. Mortar boards are available in plastic or plywood, and they average about 3 feet on a side. The plywood board needs to be wetted thoroughly before mortar is put on it to keep the board from absorbing moisture and drying the mortar.

A smaller version of the mortar board is the hawk, shown in *Figure 36*. This small board carrier with a pole-grip handle underneath is used to carry small amounts of mortar. A close relative of the plasterer's hawk, it is useful for tasks such as pointing. A wood hawk needs to be wetted thoroughly before mortar is put on it.

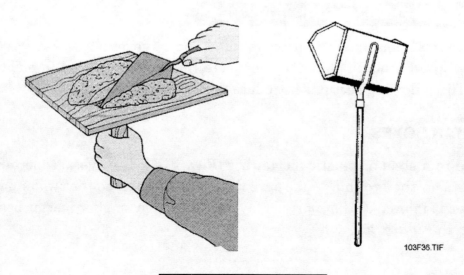

103F36.TIF

Figure 36. Hawk And Hod

The hod, also in *Figure 36*, carries mortar from the mixer to the mason. It is practical for moving material in tight spaces or on scaffolding. It is a rectangular box with a very long pole handle.

An important part of the equipment needed to prepare mortar is the cubic-foot measuring box. As shown in *Figure 37*, this box measures one cubic foot of sand or cement. It is used to measure ingredients by volume in order to proportion the mortar correctly.

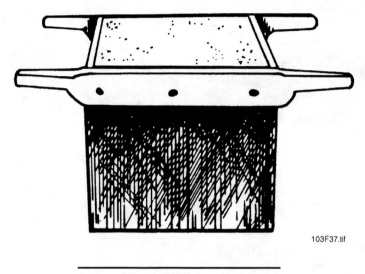

103F37.tif

Figure 37. Cubic-Foot Measure

Mortar containers and carriers should be cleaned out with water between loads of mortar. The mixing pan should be cleaned out after each batch and cleaned especially well at the end of each day.

4.2.0   MIXING ACCESSORIES

Several tools are used in mixing mortar. A long-handled shovel, *Figure 38*, is used to measure, shovel, and mix dry ingredients in the mortar box. A large hoe is used to mix wet mortar and to **temper** mortar. The mortar hoe, *Figure 38*, has a 10" blade with two holes in the blade to make it easier to pull the hoe through the mix. A square-end, short-handled shovel, *Figure 38*, is used to move mortar from the mixing box to a pan or a stand, or into a hod.

Remember to clean mixing tools with a stiff brush and water immediately after use.

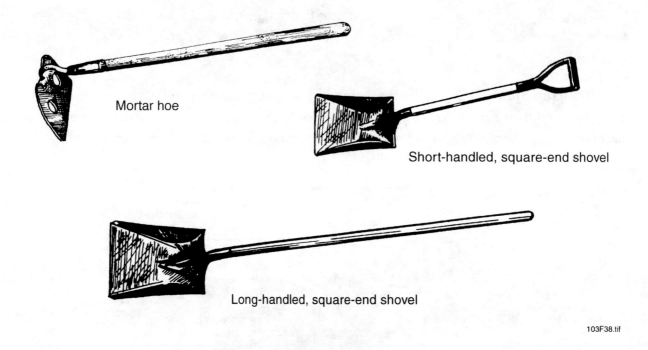

Mortar hoe

Short-handled, square-end shovel

Long-handled, square-end shovel

103F38.tif

Figure 38. Mixing Aids

4.3.0 WATER BARREL OR BUCKET

The water bucket and barrel, *Figure 39*, are the mason's friends. Made of steel, plastic, or galvanized metal, each is a necessary piece of equipment. Half-filled with water, the bucket provides an easy bath to clean mortar off hand tools. The barrel provides a bath for larger tools as well as an instant supply of water for **retempering** mortar. The bucket can also serve as a carrier for small amounts of mortar or grout.

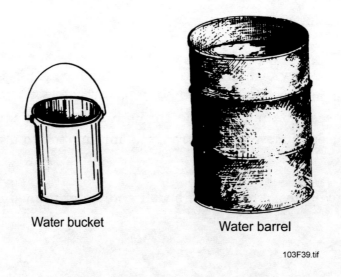

Water bucket

Water barrel

103F39.tif

Figure 39. Water Bucket And Barrel

Be sure to wet the inside surface of the bucket before filling it with mortar or grout. Always wash empty buckets out immediately to prevent mortar hardening inside. Wash the barrel out at the end of the day.

4.4.0 BARROWS

Masons use two types of barrows, shown in *Figure 40*. The standard contractor's wheelbarrow is made of steel with wooden handles. It can carry about 5 cubic feet of masonry or mortar. The second type of barrow is a pallet on wheels. Its wooden body has no sides, so it is convenient to unload masonry units from either side. However, it must be unloaded so that it does not become unbalanced and tip over. This brick barrow works well for moving bags of cement, piles of block, and other bulky materials.

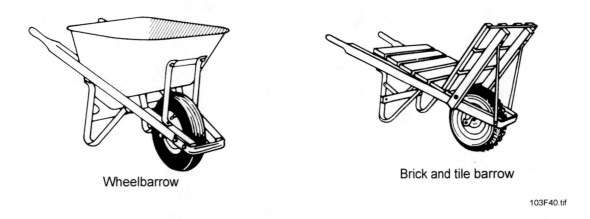

Wheelbarrow

Brick and tile barrow

103F40.tif

Figure 40. Barrows For Masonry Units

Both types of barrows need large, air-filled tires to move over bumpy ground without tipping. Check the wheels before using any barrow.

5.0.0 POWER TOOLS

Power tools bring time and labor savings to masonry work. Most mechanical and power tools have been developed since the 1920s and have become an integral part of masonry.

When using power tools and equipment, always follow power tool safety rules. Inspect items before using to make sure they are clean and functional. Disconnect power cords or turn off engines before inspecting or repairing power equipment and before the final cleaning at the end of the day.

5.1.0 SAWS

Masonry saws make more accurate cuts than a mason's hammer or brick set. The saw does not weaken or fracture the material as hand tools do. Masonry saws are available as hand-held or table-mounted models. Some larger table saws can be operated with foot controls, which leaves both hands free for guiding the piece to be cut.

Figure 41 shows a large masonry saw. Note that the saw is used with a carrier or conveyor tray on tracks. The masonry unit to be cut is placed on the carrier tray, aligned, and carried against the blade.

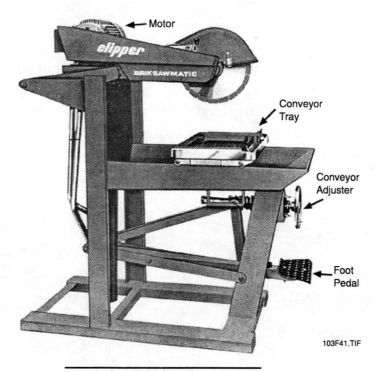

Figure 41. Large Masonry Saw

A smaller version of the masonry saw is portable. It is easily moved about and quickly set up at different areas of the job site. As shown in *Figure 42*, this saw also has a carrier or conveyor tray that runs on tracks. This saw has hand controls, and is also known as a chop saw because the blade is lowered (chopped) onto the brick.

Figure 42. Small Masonry Saw

Masonry table saws use diamond, carborundum, or other abrasive blades. Diamond blades must be irrigated to prevent them from overheating and burning up. The irrigating water wets the masonry unit, which must be allowed to dry out before it can be laid. The dry abrasive blades cut slowly and cannot cut as thinly, but the masonry can go directly to the mortar bed. Dry cutting produces dust which must be vented away from workers and the work site.

Hand-held masonry saws with 14" blades (shown in *Figure 43*) are also known as rapid cut or quickie saws. Hand-held saws with 12" blades are sometimes called cut-off saws. These smaller saws can be powered by gasoline, electricity, or hydraulics. The diameter of a blade on a hand-held saw is usually up to 14", about half the blade diameter on a table saw. Using a hand-held saw calls for extra caution. You may need to build a simple jig to keep the masonry units in place as they are cut.

103F43.tif

Figure 43. Hand-Held Masonry Saw

When using power saws of any size, review saw control operations before starting to cut.

CAUTION: Make sure electrical saws are grounded.

- Wear a hard hat and eye protection to guard against flying chips.
- Wear rubber boots and gloves to reduce the chance of electric shock.
- Wear a respirator when using a dry-cut saw.
- Check that the blade is properly mounted and tightened before starting to cut.
- Use the conveyor cart or a backup block for saws without a conveyor.

5.2.0 SPLITTERS

Masonry splitters are mechanical cutters for all types of masonry units. They do not have engines but use hydraulic power. Splitters range from small hand-operated units, as shown in *Figure 44*, to massive foot-operated hydraulic splitters. These mechanical units offer more precision than a brick hammer or brick set and are faster, especially when there are many bricks to be cut.

103F44.tif

Figure 44.   Mechanical Masonry Splitter

Unlike saws, splitters do not create large volumes of dust and do not have high-speed blades. Splitters do make cuts as neatly as saws. Larger units, as shown in *Figure 45*, can deliver as much as 19 tons of cutting pressure but are precise enough to shave ¼ of an inch off a brick or stone. Large hydraulic splitters can accommodate larger masonry units than most large saws.

301F45.tif

Figure 45. Foot-Operated Hydraulic Splitter

5.3.0 GRINDERS

The tuckpoint grinder, *Figure 46*, is a hand-held, electric grinding tool designed to grind out old bed and head joints in a masonry wall. It has shatterproof blades and a safety guard on the part of the machine facing the mason.

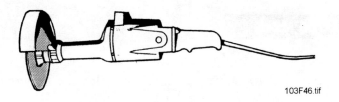

103F46.tif

Figure 46. Tuckpoint Grinder

| *WARNING!* | Wear a helmet and eye protection when using a tuckpoint grinder. |

5.4.0 POWER DRILLS AND POWDER DRIVERS

Usually, masons set bolts to anchor metalwork or wood when the mortar is still pliable. When the mortar is hardened, a power drill or powder fastener driver is needed. Drilling in masonry takes a combination of power and speed, so a ¼-inch drill is not adequate. Enough power is delivered by a ⅜-inch or ½-inch drill with a carbide-tipped bit.

Powder fastener drivers are usually designed to drive a proprietary line of bolts, anchors, and other types of fastening. These fastener drivers use a small explosive charge to drive a pin or stud into masonry. Specially hardened pins or studs are used, and different charge sizes are available. Some models use compressed air. *Figure 47* shows a powder driver used to install anchors into masonry.

103F47.tif

Figure 47. Powder Driver

Operators of powder drivers must be licensed or certified. Do not operate a powder driver without the proper training and documentation.

CAUTION: Wear ear protection when using any kind of powder driver. Also wear eye protection to guard against misfires or flying mortar.

6.0.0   POWER EQUIPMENT

Power equipment, like power tools, brings speed and economy to the masonry building process. When using power equipment, follow the rules for power tool safety.

6.1.0   MORTAR MIXER

On most commercial jobs, mortar is mixed in a powered mortar mixer or pug mill. The mixer has an electric or gasoline engine, and is usually on a set of wheels as shown in *Figure 48*. The mixer portion consists of a drum with a turning horizontal shaft inside it. Blades are attached to the shaft and revolve through the mix. The dump handle and drum release allow emptying the mortar to a pan or board.

Figure 48. Mortar Mixer

Mixers range in size from 1 to 7 cubic feet with the typical mixer holding about 4 cubic feet. The mixer drum needs to be washed out immediately after each use to keep mortar from hardening inside it.

6.2.0 SPREADER

The mortar spreader, *Figure 49*, is seen frequently in Europe but is not common in the United States. This spreader is a metal bucket equipped with a vibrator and electric motor. It drops two parallel, continuous lines of measured mortar along the horizontal edge of a course of block. Metal flanges or lips on each side of the spreader fit over the course and keep the spreader properly positioned as it moves along.

Figure 49. Powered Mortar Spreader

This spreader is used primarily on larger construction jobs with a number of long, straight, block walls. The spreader will lay the face shell bedding mortar as it moves along.

6.3.0 MASONRY PUMP AND VIBRATOR

The masonry pump is used to deliver mortar or grout to a high location. Grout is usually pumped when it is used to fill the cores in a block wall. *Figure 50* shows ready-mix grout which has been moved from the mixer to the intake hopper of the grout pump to the delivery hose.

Figure 50. Grout Pumping

At the deposit site, the mason guides the grout into the cores. After the grout is delivered, it is sometimes vibrated to eliminate air holes. *Figure 51* shows a hand-held, electric grout vibrator. The snake-like end of the vibrator is inserted into the core.

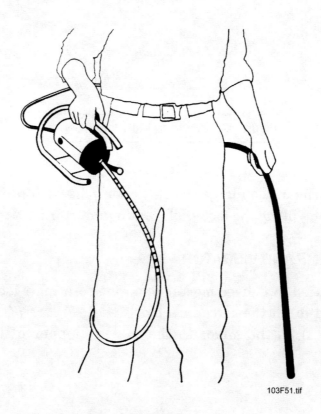

Figure 51. Hand-Held Vibrator

The mason inserts the vibrator into each core to make sure that air pockets are removed and the grout is consolidated. Steel reinforcement may also be placed in the cores before the grout is added.

6.4.0 PRESSURIZED CLEANING EQUIPMENT

Pressurized cleaning equipment has effectively replaced bucket-and-brush cleaning for masonry structures. Pressurized cleaning uses some type of abrasive material under pressure to scour the face of the masonry. The generic equipment for pressurized cleaning is an air compressor, a tank or reservoir for pressurizing, a delivery hose, and a nozzle or tip. *Figure 52* shows this type of equipment. There are three types of pressurized cleaning systems, differing mostly in the abrading material they deliver. The following sections describe pressure washing, steam cleaning, and sandblasting.

103F52.TIF

Figure 52. Pressurized Cleaning Equipment

6.4.1 Pressure Washing

Pressure washing can be the most gentle method for cleaning masonry structures. Also called high-pressure water cleaning, this is a newer cleaning technique. The pressure washer uses a compressed air pump to pressurize water and to deliver it in a focused, tightly-controlled area. Sometimes pressure washing is done after manual cleaning.

Pressure washing seems to have the best results when the operator uses a fan-type tip, dispersing the water through 25 to 50 degrees of arc. The amount or volume of the water has more effect than the amount of the pressure. The minimum flow should be 4 to 6 gallons per minute. Usually, the compressor should develop from 400 to 800 psi water pressure for the most effective washing. It is important to keep the water stream moving to avoid damaging the wall.

Pressure washing can be used in combination with various proprietary cleaning compounds. The newest pressure washers have boilers to deliver hot water. Training and practice are necessary to properly control the mix of chemicals, pressure, and spray pattern.

6.4.2   Steam Cleaning

Steam cleaning is another method of cleaning masonry. The steam cleaning equipment is very similar to the pressure washer, with the addition of a boiler to heat the water. Steam cleaning requires much less water and is used extensively on interiors and ornately carved stonework. Steam can be used in combination with chemicals to remove applied coatings such as paint from masonry.

Steam cleaning is dangerous because the steam is not only scalding but under pressure. The steam also obscures the vision of the equipment operator. Steam cleaning is a highly specialized field, and this work is usually done by a trained operator.

6.4.3   Sandblasting

Sandblasting is the oldest method of pressurized cleaning. It has the most capability of damaging or scarring the brick face, so it is best done by a trained operator.

Abrasives used include wet or dry grit, round or sharp-grained sand, crushed nut shells, rice hulls, egg shells, silica flour, ground corncobs, and other softer abrasives.

7.0.0   LIFTING EQUIPMENT

Once the mason's work reaches higher than about 4 feet, it is not efficient to work standing on the ground. Platforms or some type of scaffolding lifts the work level of the mason and the masonry units. An above-grade masonry work station usually has a mortar pan or board, the mason's tools, and a stack of masonry units.

Typically, masonry units arrive at the job site bound or bundled, and palletized. Each bundled **cube** may contain about 500 standard bricks, or about 90 standard blocks, depending on the manufacturer. Depending on the job, different types of equipment will move the cubes to the work station. The following sections deal with lifting equipment for materials handling.

WARNING!   Materials movement presents the most dangerous activity on any job. Stay out of the way of moving materials.

Work safety calls for:

- Establishing clear pathways for materials movement
- Using a consistent set of signals to alert workers to materials movement
- Refusing to ride on materials or on equipment as it moves materials
- Staying out of the area between the materials movement path and any wall or heavy equipment

7.1.0 PULLEY HOISTS

The oldest and simplest form of materials movement technology is a pulley. A pulley is also known as a block and tackle. It is a rope passed around two gears that mechanically increase the leverage or lifting force of the rope. The simple pulley does not have any safety features, such as a line brake, so it is not recommended for professional work.

Adding safety features and a ratcheting lever to increase lift results in the hand-operated ratchet pulley. Shown in *Figure 53*, this pulley hoist can raise a cube of masonry to about 12 feet. This pulley has a gear or pawl, which acts as a ratchet. The top hook attaches to a hoist arm for support, and the lower hook attaches to the load to be lifted.

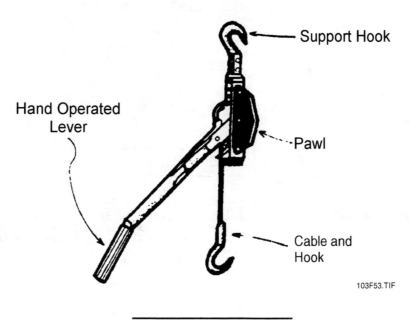

Support Hook

Hand Operated Lever

Pawl

Cable and Hook

103F53.TIF

Figure 53. Ratchet Pulley

A more sophisticated version of the pulley is found on the ladder platform hoist (*Figure 54*). This small hoist has a gasoline or electric engine, a pulley system, a take-up reel for lifting cable, a ladder, a lifting trolley, and hand controls. The combination is lightweight enough for one person to set up.

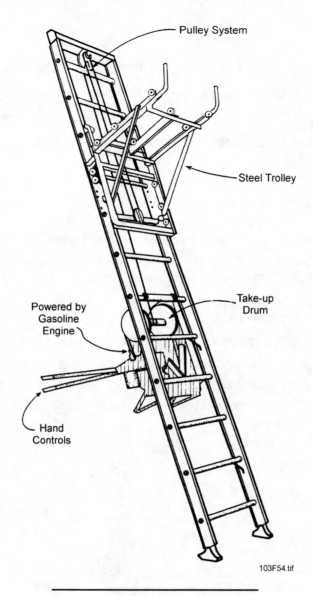

Figure 54. Ladder Platform Hoist

The typical ladder hoist can lift 400 pounds to a height of 16 to 40 feet. A plywood board can be put over the steel trolley for raising nonpalletized materials. Some hoists come with gravel or mortar hoppers that fit on the trolley. The trolley is raised by the pulley and uses the sides of the ladder as rails.

7.2.0   MATERIALS HOIST

A more complex version of the pulley hoist is the materials hoist. The portable materials hoist, shown in *Figure 55*, can be mounted on a truck bed for easy movement around the job site. The materials hoist includes a lift platform, lift cabling, and a gasoline, diesel, or electric motor. There is usually a pulley system as well (not visible in this illustration). The lift platform may have a cage around it. Materials hoists can lift from 1,000 to 5,000 pounds over a vertical distance of up to 300 feet. The larger hoists are usually not portable but attached to the side of the structure being built.

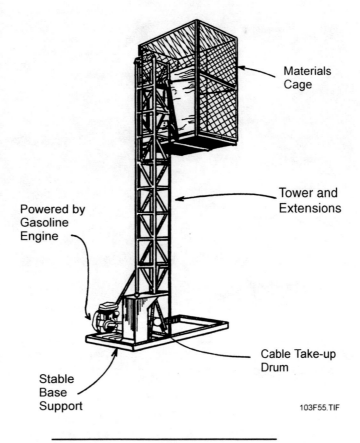

Figure 55. Portable Materials Hoist

Materials hoists are not for lifting persons. Personnel hoists have guardrails, doors, safety brakes, and hand controls in addition to the features of the materials hoist. Personnel hoists can be used for materials, but material hoists cannot be used for personnel.

WARNING!   Do not ride on a materials hoist. Do not use the materials hoist as a work platform. A materials hoist has no safety features or brakes.

7.3.0   HYDRAULIC LIFT MATERIALS TRUCK

The hydraulic lift materials truck, *Figure 56*, has a hydraulic boom arm for unloading masonry and other materials. The truck operator lowers stabilizing arms to prevent the truck overturning. The hydraulic lift arm attaches to the load and swings out to unload each cube of masonry. If the masonry is not palletized, the cubes have openings in the bottom to accommodate the lift arm attachment prongs.

103F56.TIF

Figure 56. Hydraulic Lift Materials Truck

The hydraulic lift materials truck generally is not used for lifting materials above ground level, but is used to quickly unload many cubes close to the work site or into a stockpile.

7.4.0   CRANES AND DERRICKS

Larger and more versatile versions of the boom arm mechanism can be hydraulic or operate by cabling. This class of equipment includes cranes and derricks. Some of these use gasoline-powered motorized cables with hooks for raising and lowering heavy weights, and for moving them over a horizontal distance as well. The cabling runs through pulleys and a boom arm which also moves. There are several different types of cranes used to move masonry units on large construction projects:

- Tower cranes stand alongside, or in the middle of, the building under construction.
- Mobile cranes are mounted on crawler tracks or truck beds and move about the job site.
- Conventional cranes and derricks stand away from the building by a distance according to the length of the boom arm.

WARNING!	Cranes are a necessary part of large-scale construction projects and may generate hazards. In high-rise construction, materials movement by crane poses the greatest safety hazard to workers on the job.

Work safety calls for:

- Establishing clear pathways for materials movement
- Using a consistent set of signals to alert workers to movement
- Assigning individual responsibilities for each worker during materials movement

7.4.1 Tower Cranes

Tower cranes, as shown in *Figure 57*, have a vertical tower and horizontal swinging boom or jib. The boom can move in a full-circle horizontal swing unobstructed by the tower. The crane has a counterweight on the end of the boom to balance the load, and a cab on the tower for the operator.

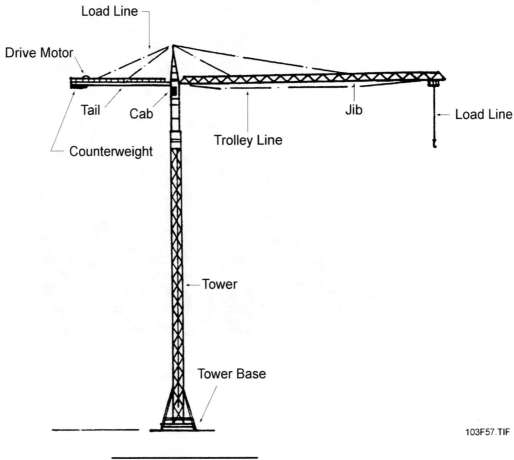

103F57.TIF

Figure 57. Tower Crane

Tower cranes are the most popular because of their freedom of swing. Because they are self-balancing, they do not need to be tied to the structure. These cranes can be mounted in the middle of a high-rise building under construction and moved from one level to the next as the structure rises.

7.4.2 Mobile Cranes

Mobile cranes lift with swinging booms raised and lowered by a boom hoist cable, as shown in *Figure 58*. This particular mobile crane has a hydraulic boom arm that extends its reach. The boom and operator's cab can be mounted on a truck bed or crawler tracks. The counterweight is usually at the bottom end of the boom or under the operator's cab. The main advantage of this crane is that it can move about the site.

103F58.TIF

Figure 58. Mobile Crane On Truck Bed

7.5.0 FORKLIFTS, PALLET JACKS, AND BUGGIES

Forklift tractors, or forklifts, have hydraulic lifting arms that move up and down. The lifting arms cannot move forward, so the equipment does. They are called forklifts because of the fork shape of the prongs on the lifting arm. The prongs fit into the holes in a pallet or the bottom of a cube of masonry. These mobile lifters are used to move masonry from a stockpile to a workstation. If less than a full cube of masonry is needed, the materials must be stacked on a pallet for raising.

The forklift consists of a gasoline or electric engine with a seat for the operator and a hydraulic powered lift. Because of their size, they are not used for raising cubes above the first floor.

Mason's forklifts or pallet jacks are specialized for masonry handling. They carry material from a stockpile to a workstation and can lift as high as one scaffold height. As shown in *Figure 59*, the pallet jack has a gasoline, diesel, or electric engine, with a hydraulic forklift. There is no seat for the operator, who stands. The load capacity of a typical pallet jack varies from ½ to 1 cube of masonry.

103F59.TIF

Figure 59. Mason's Forklift Or Pallet Jack

Motorized buggies carry mortar and masonry units around the job site. As shown in *Figure 60*, they are large, motorized wheelbarrows with two wheels in front and one wheel in back. The bin dumps its load mechanically. There is a shelf in the back for the operator to stand on as the buggy moves.

103F60.TIF

Figure 60. Power Buggy

7.6.0 CONVEYORS

Conveyors move materials from ground level to medium heights. They can be linked by swivels to carry materials over long distances. At an angle of 60 degrees, a typical conveyor can carry a load of 500 pounds at speeds of about 72 feet per minute; this material is moving fast.

CAUTION: Keep clear of the discharge end of a conveyor belt. Check the materials discharge arrangement before the conveyor starts.

Each span of a conveyor is called a flight. Mechanically, a conveyor flight has a continuous belt, moved by a chain or hydraulic drive, and powered by a gasoline or electric engine. The belt is typically about 16" wide. Specialized masonry conveyor belts have cleats, as shown in *Figure 61*, to keep the masonry from slipping down the belt. Other types of belts have troughs or bins to carry mortar or mortar ingredients.

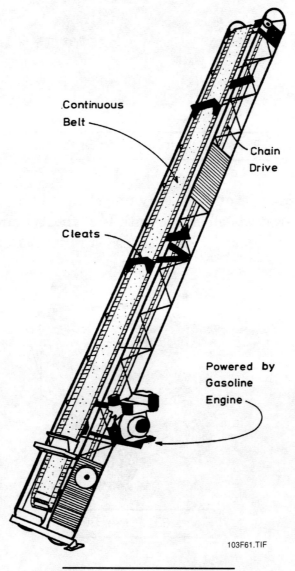

103F61.TIF

Figure 61. Conveyor Flight

8.0.0 SCAFFOLDING

There are very few masonry jobs that do not require some type of scaffolding. As the working level rises above 4 feet, masons cannot work as efficiently. It is cost- and time-effective to raise the mason and the supply close to the work. Masons themselves sometimes erect the scaffolding they work on because they are concerned with its safety and stability.

There are several types of scaffolding in common use, ranging from the simple to the complex and more costly. All scaffolding consists of vertical members that support horizontal members above the ground. Additional parts are guardrails and toeboards for the horizontal members, and base or foundation plates for the vertical members. All scaffolding must be erected, moved, and disassembled under the supervision of a competent person. All scaffolding must be assembled according to federal safety regulations and attached to the building structure at minimum distances of 25 feet. The following sections describe some of the most commonly-used scaffold systems.

8.1.0 TRESTLE

Trestle support scaffolding is the oldest type. Today, this type of scaffolding is made from tubular steel supports similar to sawhorses. The platforms can be made of planking as shown in *Figure 62*.

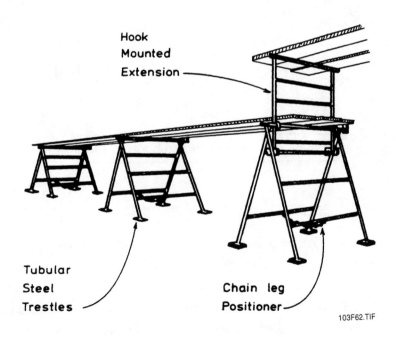

103F62.TIF

Figure 62. Tubular Steel Trestles

Tubular steel trestles come in several sizes, with many advantages over the old system of wood trestles. They are lightweight, strong, and easy to move around. Their legs have baseplates to keep them from sinking into the ground, and chains to limit leg movement. Their ladder sides offer the ability to mount planks at different heights. They can take short vertical extensions with hook-mounting devices. They will not rot or split as wood does.

Regular dimensional lumber planking can be used for the platforms on steel trestles, or hook-mount scaffold planks can be used. As shown in *Figure 63*, these hook-mount planks are of plywood, with steel rails with hooks at the ends. These hook-mount planks are also used with tubular steel sectional scaffolding.

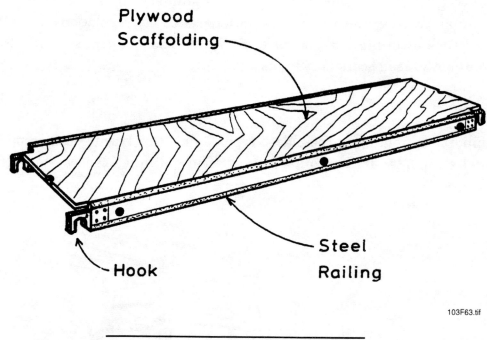

103F63.tif

Figure 63. Hook-Mounted Scaffold Plank

8.2.0 TUBULAR STEEL SECTIONAL SCAFFOLDING

Tubular steel sectional scaffolding is usually found on jobs of more than one story. Its strength, light weight, durability, and ease of erection have allowed steel scaffolding to replace wood almost completely in commercial use. The steel frame sections come in several heights, with a typical 4- or 5-foot width. As shown in *Figure 64*, two commonly used frame sections are the ladder-braced and the walk-through frames. The walk-through allows the mason an easy passage along the scaffold planks.

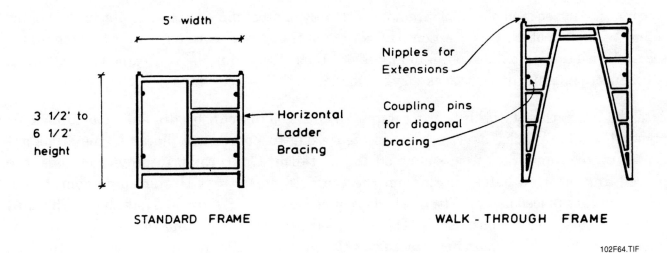

5' width

3 1/2' to
6 1/2'
height

Horizontal
Ladder
Bracing

STANDARD FRAME

Nipples for
Extensions

Coupling pins
for diagonal
bracing

WALK - THROUGH FRAME

102F64.TIF

Figure 64. Scaffold Frame Types

As tubular steel sectional scaffolding is set up, each frame is connected to the next frame at ground level with a horizontal diagonal brace. In addition, frames are connected at upper levels by overlapping diagonal braces. These overlapping braces must be secured to the frames by a wing nut or bolt, or by a lock device that fits over the frame coupling pin, shown in *Figure 65*.

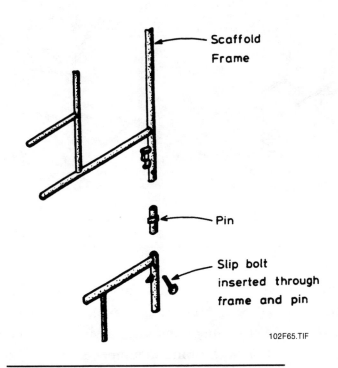

Scaffold
Frame

Pin

Slip bolt
inserted through
frame and pin

102F65.TIF

Figure 65. Connecting Scaffolding Sections

To connect each section of scaffold framing vertically, a steel pin or nipple is put in the hollow tubing at the top of the lower section. The bottom of the upper section fits over this pin and is secured with a slip bolt, as shown in *Figure 65*. Each level must be braced as it is installed. All connections must be secured as they are made.

After bracing, the platform flooring is put in place. This can be dimensional lumber or plywood scaffold planks, as was shown in *Figure 63*. Toeboards are placed as materials are loaded to keep materials from sliding off the platform. Guardrails, to protect masons, are placed and secured before work from the scaffold can start. Under some conditions, roofboards are placed as well. The finished product looks like *Figure 66*. Note that a sill board has been put under the base plates. This is a good practice for ground that is soft, damp, or liable to shift. With the materials stacked on the planking, the mason can stand on the side bracket close to the wall.

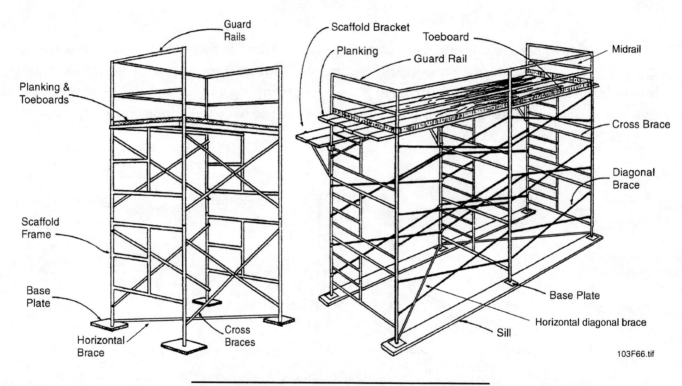

Figure 66. Completed Tubular Steel Scaffolding

Tubular steel framing is very versatile and can be configured many different ways with different accessories. Mason's side brackets, or extenders, extend the working area. Screw jacks with base plates level the framework after it is assembled. Wheels are available and can be inserted into the bottom of framing members so that the structure becomes a rolling scaffold. Rolling scaffolds should be used only for pointing and striking joints, not for laying units. Even with wheel locks, they tend to move when heavily loaded with masonry units.

8.3.0 PUTLOG

Sometimes called a "bridge," a putlog is a wooden beam supporting scaffolding. Putlogs are used where there is no solid base on which to set the scaffolding frame. Because of grade conditions at the job site, one or more putlogs may be needed to stabilize scaffolding. Putlogs are set with their greater dimension vertical and must extend at least 3 inches past the edge of the scaffold frame.

When putlogs are required, holes are left in the first few courses of a wall. The putlogs are inserted in these holes with their other ends resting on solid ground (*Figure 67*) or on a constructed base.

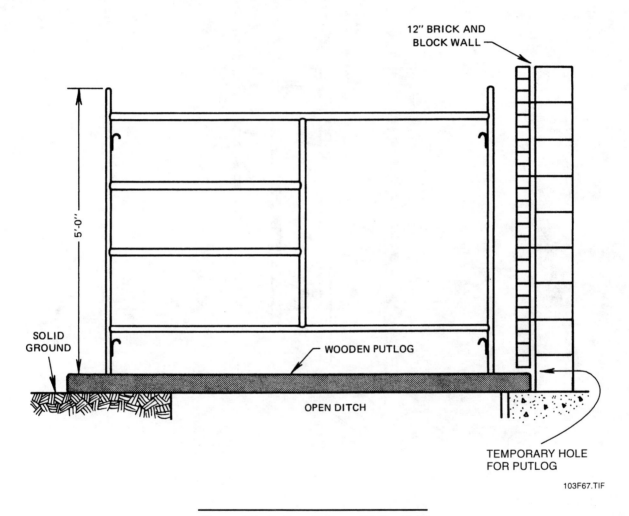

Figure 67. Putlog At Base Of Scaffold

The scaffold frame is erected on the putlogs, which provide a solid footing. The scaffolding is braced as usual and the base plates may be fastened to the putlogs. After the scaffold is disassembled, the putlogs are withdrawn from the holes and the holes are filled with masonry units.

8.4.0 STEEL TOWER

The adjustable steel tower scaffold or self-climbing scaffold (*Figure 68*) is built of vertical members braced or tied to the structure itself. The ties must be at, or closer than, every 30 feet. The scaffolding framing is attached to the vertical members. Metal planking is used for a working platform which has guardrail and toeboard safety features, and a winch mechanism. The working platform can be raised as the work progresses.

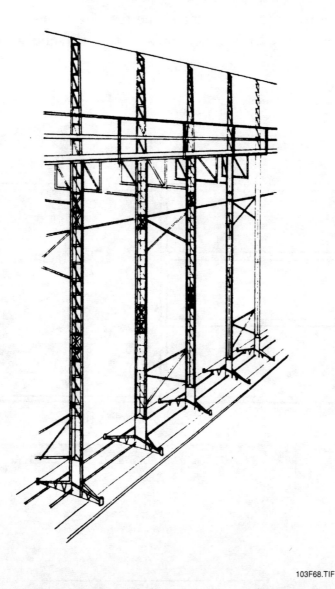

103F68.TIF

Figure 68. Adjustable Steel Tower Scaffolding

Because it requires attachment to the building, the tower scaffolding may be put up by specialized workers.

8.5.0   SWING STAGE

Swing stage scaffolding is used on tall, multistory buildings. Steel beams are fastened to the roof and steel cables are dropped to the ground. A steel cage is suspended from the cables with hangers and a planking floor is added over the frame. Guardrails, toeboards, and an overhead canopy of plywood or metal mesh complete the cage, as seen in *Figure 69*. A winch moves the cage up and down so it is always at the level of the work.

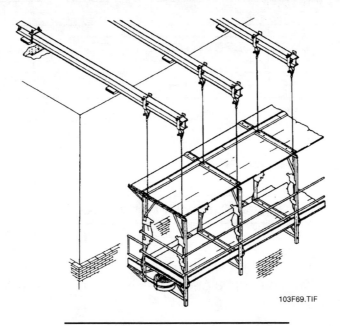

103F69.TIF

Figure 69. Swing Stage Scaffolding

Unlike the other types of scaffolding, the swing stage is usually erected by specialists. This type of scaffolding is also the safest, because the mason is completely enclosed by steel, and there are redundant backup systems on the cabling and brakes.

8.6.0   HYDRAULIC PERSONNEL LIFT

On some jobs, a hydraulic personnel lift is used as scaffolding. As discussed earlier, a hydraulic materials lift does not have safety features. Hydraulic personnel lifts have guardrails, toeboards, hand controls, and stabilizer pads. They have backup safety systems, free-fall brakes, and safety ladders. Do not use a materials lift as a scaffold.

8.7.0   FOOT OR BLOCKING BOARD

Sometimes it is necessary to raise a short platform so that the mason can reach the area below the first level of any scaffold. Typically, this is a wooden plank on several blocks or bricks. Federal scaffold safety regulations apply to scaffolds 6 feet above ground at construction sites, so they do not cover foot boards. Technically not a scaffold, this foot or blocking board is useful, but it must be safe as well. As shown in *Figure 70*, no more than two blocks should be used to support the foot board. The blocks must be placed with their face shells perpendicular to the ground and their cores aligned.

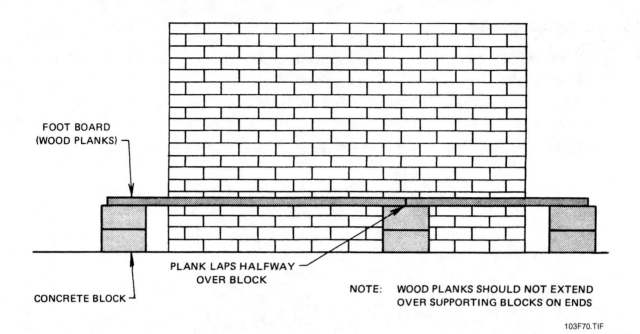

FOOT BOARD
(WOOD PLANKS)

PLANK LAPS HALFWAY
OVER BLOCK

CONCRETE BLOCK

NOTE: WOOD PLANKS SHOULD NOT EXTEND
OVER SUPPORTING BLOCKS ON ENDS

103F70.TIF

Figure 70. Foot Or Blocking Board

To avoid safety hazards, do not build a foot board more than 16 inches high. Also, do not extend the boards over the outside edges of the block supports, as they will get knocked off.

8.8.0 SCAFFOLD SAFETY

Masons spend a great deal of time working from scaffolding. The chance of an accident occurring because of poor work practice is greater on scaffolding than at any other time or place. Tubular steel scaffolding is less likely to give way than the old-fashioned wood scaffolding, but requires more care in its assembly and use.

The new federal safety rules focusing on worker safety are summarized below. These rules are only reminders for the safe assembly and use of scaffolding.

- Platforms on all working levels must be fully decked between the front uprights and the guardrail supports.

- The space between planks and the platform and uprights can be no more than 1" wide.

- Platforms and walkways must be at least 18" wide; the ladder jack, top plate bracket, and pump jack scaffolds must be at least 12" wide.

- For every 4 feet a scaffold is high, it must be at least 1 foot wide. If it is not, it must be protected from tipping by tying, bracing, or guying per the OSHA rules.

- Supported scaffolds must sit on base plates and mud sills or other steady foundations.

- The inboard ends of suspension scaffold outriggers must be stabilized by bolts or other direct connections to the floor or roof deck, or stabilized by counterweights.

- Access to and between scaffold platforms more than 2 feet above or below the point of access must be made by:
  - Portable ladders, hook-on ladders, attachable ladders, scaffold stairways, stairway-type ladders (such as ladder stands), ramps, walkways, integral prefabricated scaffold access, or equivalent means; or
  - By direct access from another scaffold, structure, personnel hoist, or similar surface.
- Never use crossbraces to gain access to a scaffold platform.

The new OSHA rules for scaffolding systems require that all workers who use or work around scaffolding must be trained to recognize the hazards of these systems, including proper erection and use. All employees who work on scaffolding must be trained by a person qualified in scaffolding hazards and uses, so they understand the procedures to control or minimize any hazards. All personnel who erect, take down, move, operate, maintain, repair, or inspect scaffolding must be trained by a competent person.

SUMMARY

This module describes a mason's hand tools: trowels, hammers, chisels, jointers, and other tools, and what they are used for. The course layout, alignment, and measuring tools are listed in detail as they are used constantly during a workday to check that a masonry structure is not out of wind. Measures and measuring tools include levels or plumb rules, squares, mason's line and fasteners, corner poles, chalk line, and two types of mason's spacing rules. Mortar mixing equipment includes water buckets and barrels, hoes, shovels, and mortar boards, boxes, and pans. Mortar boxes, boards, and pans must be cleaned after use, as should hand tools and power equipment.

Power tools and equipment make the mason's job easier. Mortar is quickly mixed in a power mixer. Power saws and hydraulic splitters are used to cut masonry units. Masonry needs to be lifted to the work location by hoists, cranes, forklifts, pallet jacks, and conveyors. Masons need to be lifted to the work location by scaffolding. Masons usually erect their own scaffolding because their safety depends on it.

References

For advanced study of topics covered in this task module, the following books are suggested:

Building Block Walls—A Basic Guide, National Concrete Masonry Association, Herndon, VA, 1988.

Masonry Design and Detailing—For Architects, Engineers and Contractors, Fourth Edition, Christine Beall, McGraw-Hill, New York, NY, 1997.

Masonry Skills, Third Edition, Richard T. Kreh, Delmar Publishers, Inc., Albany, NY, 1990.

The ABC's of Concrete Masonry Construction, Videotape, 13:34 minutes, Portland Cement Association, Skokie, IL, 1980.

ACKNOWLEDGMENTS

Figures 1, 5-10, 12-14, 16-17, 19, 22, 24, 26-29, 31-33, 35, 37-40, 44-45, 48-49, 51, 53-55, 57, 60-66 courtesy of Associated General Contractors

Figures 3, 4, 11, 18, 20, 25, and 69 courtesy of The Goodheart-Willcox Company, Inc.

Figures 21 and 30 courtesy of The Goldblatt Catalog

Figures 15, 34, and 36 courtesy of Creative Homeowner Press

Figures 2, 41, 67, and 70 courtesy of Delmar Publishers, Inc.

Figure 23 courtesy of Dover Books

Figure 47 courtesy of HILTI Corp.

Figure 68 courtesy of National Concrete Masonry Association

Figure 58 courtesy of National Crane

Figure 50 courtesy of Portland Cement Association

Figures 42-43 courtesy of Rentals Unlimited

Figures 56 and 59 courtesy of Roy Jorgensen Associates, Inc.

Figure 46 courtesy of Stanley Tools

Figure 52 courtesy of Professional Equipment and Supply

REVIEW/PRACTICE QUESTIONS

1. A trowel with a sharply angled heel is in the _____ pattern.

   a. Philadelphia
   b. London
   c. Tuckpoint
   d. Cleveland

2. For striking cutting chisels, the most commonly used tool is the _____.

   a. mason's mash
   b. block set
   c. blocking chisel
   d. bushhammer

3. The brick set and the bolster are _____.

   a. striking mauls
   b. cutting chisels
   c. special-purpose trowels
   d. jointers

4. A slicker is a(n) _____.

   a. adjustable grinder with shatterproof blades
   b. adjustable set screw on roller skate wheels
   c. flat jointer with different-sized ends
   d. device used to carry bricks without chipping or breaking

5. Mason's brushes are used for _____.

   a. keeping fingers clear of dust and grout
   b. compressing mortar and reducing water absorption
   c. applying grout to mortar joints
   d. site cleaning, finish cleaning, and acid cleaning

6. A standard size plumb rule runs _____.

   a. 42 to 48 inches
   b. 32 to 38 inches
   c. 18 to 24 inches
   d. 38 to 42 inches

7. Spacing rules measure _____.

   a. plumb and level
   b. course or modular spacing
   c. trigs or leads
   d. pinch and rake

8. Use a _____ to quickly mark right and 45-degree angles.

    a. folding rule
    b. modular or course spacing rule
    c. combination or T-square
    d. bevel or skewback

9. Trigs and blocks _____.

    a. hold bevels taut and in place
    b. keep mortar in place
    c. are specialized masonry units
    d. keep mason's line taut and in place

10. Box, board, pan, and hawk are _____.

    a. mortar containers
    b. grout containers
    c. masonry types
    d. joint finishing tools

11. Hydraulic splitters do not _____.

    a. cut masonry precisely
    b. accommodate large masonry units
    c. work faster than saws
    d. create large volumes of dust

12. To grind out old head and bed joints, use a _____.

    a. tuckpoint grinder
    b. hydraulic splitter
    c. bolster and maul
    d. ½-inch drill

13. A ratchet pulley is not _____.

    a. used to raise a cube 12 to 16 feet
    b. used to raise a cube 8 to 12 feet
    c. equipped with safety features
    d. equipped with a pawl

14. Materials hoists do not have _____.

    a. cages and cabling
    b. safety brakes and hand controls
    c. platforms and lift engines
    d. truck bed mounts

15. Cubes are quickly unloaded by a _____.

    a. hydraulic splitter
    b. hydraulic materials hoist
    c. hydraulic lift materials truck
    d. pressurized foot board

16. Typical conveyors can carry a load of _____.

    a. 500 pounds at about 72 feet per minute
    b. 600 pounds at about 90 feet per minute
    c. 400 pounds at about 50 feet per minute
    d. 300 pounds at about 90 feet per minute

17. A wooden beam supporting a scaffold is called a _____.

    a. trestle base
    b. screw jack
    c. tuckpointer
    d. putlog

18. Steel tower scaffolding differs from other types because it _____.

    a. has a winch mechanism
    b. is attached to the building under construction
    c. can only be used on steel-frame buildings
    d. requires roofboards

19. Swing stage scaffolding is like other types in that it _____.

    a. has no rigid vertical members
    b. hangs from the roof of the building
    c. uses a cage with a winch
    d. is erected by specialists

20. A foot board _____.

    a. supports the bottom of a scaffold
    b. should not be more than 16 inches high
    c. supports the bottom of a materials lift
    d. should not be more than 24 inches high

ANSWERS TO REVIEW/PRACTICE QUESTIONS

Answer		Section Reference
1.	b	2.1.0
2.	a	2.2.2
3.	b	2.3.0
4.	c	2.4.0
5.	d	2.5.0
6.	a	3.1.0
7.	b	3.2.0
8.	c	3.3.0
9.	d	3.4.0
10.	a	4.1.0
11.	d	5.2.0
12.	a	5.3.0
13.	a	7.1.0
14.	b	7.2.0
15.	c	7.3.0
16.	a	7.6.0
17.	d	8.3.0
18.	b	8.4.0
19.	c	8.5.0
20.	b	8.7.0

The NCCER makes every effort to keep these manuals up-to-date and free of technical errors. We appreciate your help in this process. If you have an idea for improving this manual, or if you find an error, a typographical mistake, or an inaccuracy in the NCCER's Craft Training Manuals, please write us, using this form or a photocopy. Be sure to include the exact module number, page number, a description of the problem, and the correction, if possible. Your input will be brought to the attention of the Technical Review Committee. Thank you for your assistance.

Instructors – If you found that additional materials were necessary in order to teach this module effectively, please let us know so that we may include them in the Equipment/Materials list in the Instructor's Guide.

Write: Curriculum and Revision Department
National Center for Construction Education and Research
P.O. Box 141104
Gainesville, FL 32614-1104

Fax: 352-334-0932

Craft _____   Module Name _____

Module Number _____   Page Number(s) _____

Description of Problem _____

_____

_____

_____

(Optional) Correction of Problem _____

_____

_____

(Optional) Your Name and Address _____

_____

_____

Mathematics, Drawings, and Specifications

Module 28104

MATHEMATICS, DRAWINGS, AND SPECIFICATIONS

OBJECTIVES

Upon completion of this module, the trainee will be able to:

1. Understand and work with denominate numbers.
2. Read a mason's measure.
3. Convert measurements in the U.S. common system into their metric equivalents.
4. Recognize, identify, and calculate areas, circumferences, and volumes of basic geometric shapes.
5. Identify the basic parts of a set of drawings.
6. Discuss the different types of specifications used in the building industry and the sections that pertain to masonry.

Prerequisites

Successful completion of the following Task Modules is recommended before beginning study of this Task Module: Core Curricula; Masonry Level 1, Modules 28101 through 28103.

Required Trainee Materials

1. Trainee Task Module
2. Pencil and paper
3. Tape measure
4. Four-function calculator

COURSE MAP

This course map shows all of the task modules in the first level of the Masonry curricula. The suggested training order begins at the bottom and proceeds up. Skill levels increase as a trainee advances on the course map. The training order may be adjusted by the local Training Program Sponsor.

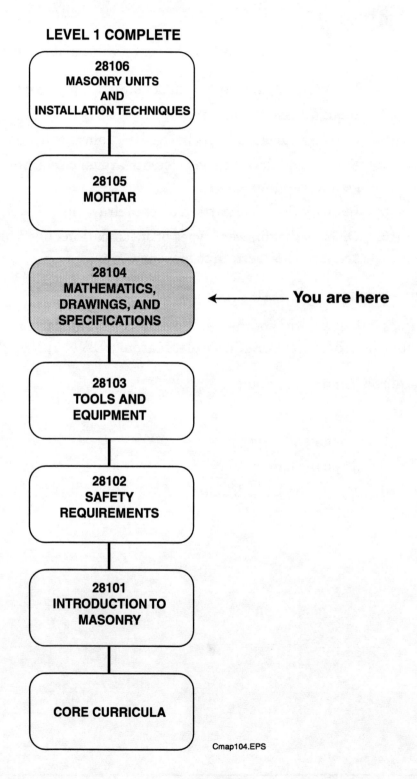

LEVEL 1 COMPLETE

28106
MASONRY UNITS
AND
INSTALLATION TECHNIQUES

28105
MORTAR

28104
MATHEMATICS,
DRAWINGS, AND
SPECIFICATIONS   ← **You are here**

28103
TOOLS AND
EQUIPMENT

28102
SAFETY
REQUIREMENTS

28101
INTRODUCTION TO
MASONRY

CORE CURRICULA

Cmap104.EPS

TABLE OF CONTENTS

Trade Terms Introduced In This Module

ACI: American Concrete Institute.

ASCE: American Society of Civil Engineers.

BOCA: Building Officials and Code Administrators.

CABO: Council of American Building Officials.

Denominate numbers: Those numbers indicating a unit of measure, such as feet or tons.

ICBO: International Conference of Building Officials.

Nominal dimension: The size of the masonry unit plus the thickness of one standard (½ inch or ⅜ inch) mortar joint; the nominal dimensions are used in laying out courses.

SI metric units: The metric-based units of measure used in most other countries.

UBC: Uniform Building Code.

U.S. customary: The units of measure commonly used in the United States such as inches, feet, yards, miles, quarts, gallons, etc.

1.0.0 INTRODUCTION

This module covers the math tools you will need in your work as a mason. It reviews three basic skills that are essential to your success:

- Calculations for masons
- Reading plans and drawings
- Reading and meeting specifications

The next sections introduce you to the topics covered in this module.

1.1.0 MASONRY MATH

Math skills are needed at every step of your work. From understanding a set of drawings to measuring and mixing mortar, math skills enable you to work quickly and efficiently. Math skills let you know how much sand to put into the mortar mix, or how many bricks will fit into an opening. Calculation replaces trial and error to save you time and energy, materials and money. This module reviews the tools needed to perform those calculations.

Denominate numbers are those that have a unit of measure associated with them, such as feet, inches, or yards. If you work in a metric environment, it is easy to convert denominate metric numbers into other metric units. If you work in a nonmetric environment, you must learn to convert these numbers. This module gives you the tools needed to convert denominate numbers.

Masons have their own set of denominate numbers, in the course rule and modular rule. This module will teach you to be comfortable with these rules and denominate numbers.

1.2.0   READING DRAWINGS

Working drawings are an on-the-job guide to what must be done. They detail the finished product and, in some cases, the intermediate steps to get there. This module introduces basic working drawings and how to read them. A set of residential plans is discussed in detail. This module will provide you with a working knowledge of how to read drawings and plans.

1.3.0   APPLYING SPECIFICATIONS

Specifications form part of the contract between the builder and the client. The specifications contain detailed descriptions of the finished product. If there is a conflict between the plans and the specifications, the specifications are followed. This module describes specifications and their development. You will use specifications to flesh out plans and drawings.

This module assumes you are familiar with addition, subtraction, multiplication, and division of whole numbers, decimals, and fractions. You should also be able to convert fractions. If you are not comfortable with these operations, review the Core Curricula *Basic Math* module. Complete your review before beginning this module.

2.0.0   MASONRY MATH

Mathematics is used by everyone to one degree or another. For a skilled mason or other craftworker, the ability to properly and accurately solve math problems is essential. This module provides a refresher in some of the basic math concepts needed to do your work as a mason as well as providing information about calculations used specifically for masonry work.

2.1.0   DENOMINATE NUMBERS

Denominate numbers have a unit associated with them. They measure things such as feet, pounds, meters, kilos, and so on. In the United States, we use a traditional set of denominate numbers based loosely on English measures. Other parts of the world use the metric system. Here we deal with inches, feet, yards, ounces, pints, quarts, and gallons. We measure masonry units in cubic yards or tons. The unit name denominates, or identifies, what is to be added, subtracted, multiplied, or divided.

Denominate number problems take two steps to solve:

- There is the simple arithmetic operation to solve the problem.
- There is the conversion, or changing the arithmetic numbers into denominate numbers.

Converting is the process used to change inches to feet or feet to yards, or to change feet or yards back to inches. You cannot do arithmetic operations on denominate units of different measures. The different measures must be converted, just as fractions must be converted into like units in order to perform arithmetic operations. The next sections review arithmetic denominate conversions.

2.1.1 Denominate Addition

Example 1 shows the steps in solving an addition problem with denominate numbers.

Example 1   Add these denominate numbers:

	1 yard	2 feet	9 inches
+	2 yards	2 feet	5 inches

Step 1   The first step is to add the inches.

	1 yard	2 feet	9 inches
+	2 yards	2 feet	5 inches
			14 inches

Step 2   There are 12 inches to the foot. Subtract 12 from 14 inches. Place the remaining inches in the inches column and carry the foot to the foot column. The rule is to convert inches to feet when the number of inches is greater than 12.

		1 foot	
	1 yard	2 feet	9 inches
+	2 yards	2 feet	5 inches
			14 inches
			− 12 inches
			2 inches

Step 3   Now add the foot column.

		1 foot	
	1 yard	2 feet	9 inches
+	2 yards	2 feet	5 inches
		5 feet	14 inches
			− 12 inches
			2 inches

Step 4   Convert the feet to yards. Subtract 3 feet from 5 feet. Place the remaining feet in the feet column and carry the yard to the yard column. The rule is to convert feet to yards when the number of feet is greater than 3.

		1 foot	
	1 yard	2 feet	9 inches
+	2 yards	2 feet	5 inches
		5 feet	14 inches
		− 3 feet	− 12 inches
		2 feet	2 inches

Step 5   Carry the yard to the yards column and add.

	1 yard	1 foot	
	1 yard	2 feet	9 inches
+	2 yards	2 feet	5 inches
		5 feet	14 inches
		− 3 feet	− 12 inches
	4 yards	2 feet	2 inches

Again, inches must be converted to feet when they exceed 12, and feet must be converted to yards when they exceed 3.

Exercises 1 through 4 will give you practice in adding denominate numbers. You can check your work by looking up the answer to each exercise in Appendix A.

Exercise 1   Add the following:

	2 yards	1 foot	10 inches
+	3 yards	2 feet	5 inches

Exercise 2   Add the following:

	4 yards	2 feet	9 inches
+	6 yards	2 feet	4 inches

Exercise 3   Add the following:

	7 yards	2 feet	7 inches
+	4 yards	1 foot	6 inches

Exercise 4   Add the following:

	8 yards	1 foot	9 inches
+	4 yards	1 foot	7 inches

2.1.2   Denominate Subtraction

Subtraction problems call for similar conversions. Example 2 shows the steps in solving a subtraction problem with denominate numbers.

Example 2   Subtract these denominate numbers:

	5 yards	2 feet	7 inches
−	2 yards	2 feet	8 inches

Step 1   The first step is to subtract the inches. Since 7 is less than 8, borrow 12 inches from the foot column, leaving 1 foot. The 12 inches is added to the 7 inches, to get 19 inches. Now subtract the 8 inches from the 19 inches.

		1 foot	(12 + 7) = 19 inches
	5 yards	2̸ feet	7̸ inches
−	2 yards	2 feet	8 inches
			11 inches

Step 2   The next step is to subtract the foot column. Since 1 is less than 2, borrow 3 feet from the yard column, leaving 4 yards. The 3 feet is added to the 1 foot to get 4 feet. Now subtract the 2 feet from the 4 feet.

	4 yards (3 + 1)	= 4 feet	
		1̸ foot	(12 + 7) = 19 inches
	5̸ yards	2̸ feet	7̸ inches
−	2 yards	2 feet	8 inches
		2 feet	11 inches

Step 3   Now subtract the yard column. Since you borrowed 1 yard from the 5 yards, you have 4 yards left.

	4 yards (3 + 1)	= 4 feet	
		1̸ foot	(12 + 7) = 19 inches
	5̸ yards	2̸ feet	7̸ inches
−	2 yards	2 feet	8 inches
	2 yards	2 feet	11 inches

In subtraction, remember that borrowing does not give the 10 units that you get **working with** ordinary numbers. The borrowed amount depends on the denominate **measure you are** converting.

Exercises 5 through 8 will give you practice in subtracting denominate numbers.

Exercise 5   Subtract the following:

	6 yards	2 feet	6 inches
−	2 yards	1 foot	8 inches

Exercise 6   Subtract the following:

	5 yards	2 feet	5 inches
−	3 yards	2 feet	7 inches

Exercise 7   Subtract the following:

	7 yards	2 feet	8 inches
−	4 yards	1 foot	10 inches

Exercise 8   Subtract the following:

	4 yards	2 feet	4 inches
−	1 yards	1 foot	7 inches

2.1.3   Other Denominate Measures

Other denominate numbers commonly used in the United States are listed in *Figure 1.*

1 gallon	4 quarts
1 quart	2 pints
16 liquid ounces	1 pint
128 liquid ounces	1 gallon
144 square inches	1 square foot
9 square feet	1 square yard
27 cubic feet	1 cubic yard
16 liquid ounces	1 pound
2,000 pounds	1 ton

104F01.EPS

Figure 1. Common Measures

2.2.0 MASON'S DENOMINATE MEASURES

Masons have two denominate numbering systems of their own, the course system and the modular system. As noted in *Tools and Equipment*, masons have two kinds of rules for these two measures. When working with these rules (or any measuring tools) it is important to:

• familiarize yourself with the scale;

• take readings carefully, and take them twice, to avoid making costly mistakes.

The old carpentry rule applies here: "Measure twice and cut once."

2.2.1 The Course System

The course system predates the modular system of measurement. The course system is also called the brick spacing rule or system. The course counter rule numbers the courses of different sizes of brick that will fill a vertical space.

This rule measure is used to lay out and space standard brick courses to nonmodular dimensions. The rule has inches on the other side, marked in sixteenths of an inch. Nonmodular course spacing is measured and estimated with the brick spacing rule. *Figure 2* shows the standard brick spacing rule.

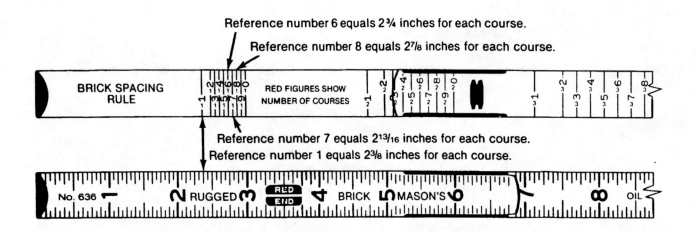

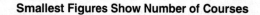

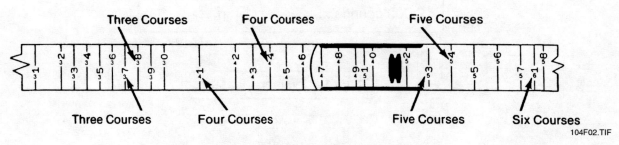

Figure 2. Reading The Standard Spacing Rule

The large black figures on the rule are references for the nominal sizes of standard brick and mortar. The small red figures are at right angles to the black figures. They count the number of courses for that size brick. Note that all the reference measures fall between 2⅜ and 3 inches. The rule has a gauge at the beginning that measures the size of one brick. It is used to identify the size of the brick so you will know which scale to read.

The standard spacing rule is useful for marking a corner pole or deadman. The first step in using the standard spacing rule is to check the specifications for the size of brick and/or the course height. Then determine the height to be filled by the courses. The third step is to transfer the correct markings from the spacing rule to the corner pole.

2.2.2 The Modular System

Today, brick is made for use on the modular grid system of building. The dimensions are based on a 4-inch unit of measure called the module. The grid system makes it easier to combine different materials in a construction job. The grid system allows the traditional measurements for different materials to intersect, so that different kinds of joinings can be measured or calculated easily.

In modular design, the **nominal dimension** of a masonry unit is the specified or manufactured dimension plus the thickness of the standard mortar joint to be used. That is, the brick size is designed so that when the size of the mortar joint is added to any of the brick dimensions (thickness, height, length) the sum will equal a multiple of the 4-inch grid.

For example, a modular brick with a nominal length of 8 inches will have a manufactured dimension of 7½ inches if it is designed to be laid with a ½-inch mortar joint. It will have a manufactured dimension of 7⅝ inches if it is designed to be laid with a ⅜-inch joint.

Figure 3 shows nominal and actual manufactured dimensions for nominal brick sizes and actual dimensions for nonmodular brick sizes. It includes the planned joint thickness of ⅜ of an inch or ½-inch. The last column shows the number of courses required for each size of brick to equal a 4-inch modular unit or a multiple of a 4-inch unit.

Most masonry materials will tie and level off together at a height of 16 inches vertically. Two courses of blocks or six courses of standard bricks, mortared, will both equal 16 inches vertically. As a result, the wythes can be tied together at 16-inch intervals or at multiples of 16 inches.

Modular Brick Sizes

Unit Designation	Nominal Dimensions Inches			Joint Thickness Inches	Specified Dimensions Inches			Vertical Coursing Inches
	w	h	l		w	h	l	
Modular	4	$2^2/_3$	8	$^3/_8$ $^1/_2$	$3^5/_8$ $3^1/_2$	$2^1/_4$ $2^1/_4$	$7^5/_8$ $7^1/_2$	3C = 8
Engineer Modular	4	$3^1/_5$	8	$^3/_8$ $^1/_2$	$3^5/_8$ $3^1/_2$	$2^3/_4$ $2^{13}/_{16}$	$7^5/_8$ $7^1/_2$	5C = 16
Closure Modular	4	4	8	$^3/_8$ $^1/_2$	$3^5/_8$ $3^1/_2$	$3^5/_8$ $3^1/_2$	$7^5/_8$ $7^1/_2$	1C = 4
Roman	4	2	12	$^3/_8$ $^1/_2$	$3^5/_8$ $3^1/_2$	$1^5/_8$ $1^1/_2$	$11^5/_8$ $11^1/_2$	2C = 4
Norman	4	$2^2/_3$	12	$^3/_8$ $^1/_2$	$3^5/_8$ $3^1/_2$	$2^1/_4$ $2^1/_4$	$11^5/_8$ $11^1/_2$	3C = 8
Engineer Norman	4	$3^1/_5$	12	$^3/_8$ $^1/_2$	$3^5/_8$ $3^1/_2$	$2^3/_4$ $2^{13}/_{16}$	$11^5/_8$ $11^1/_2$	5C = 16
Utility	4	4	12	$^3/_8$ $^1/_2$	$3^5/_8$ $3^1/_2$	$3^5/_8$ $3^1/_2$	$11^5/_8$ $11^1/_2$	1C = 4

Nonmodular Brick Sizes

Unit Designation	Nominal Dimensions Inches			Joint Thickness Inches	Specified Dimensions Inches			Vertical Coursing Inches
	w	h	l		w	h	l	
Standard				$^3/_8$ $^1/_2$	$3^5/_8$ $3^1/_2$	$2^1/_4$ $2^1/_4$	8 8	3C = 8
Engineer Standard				$^3/_8$ $^1/_2$	$3^5/_8$ $3^1/_2$	$2^3/_4$ $2^{13}/_{16}$	8 8	5C = 16
Closure Standard				$^3/_8$ $^1/_2$	$3^5/_8$ $3^1/_2$	$3^5/_8$ $3^1/_2$	8 8	1C = 4
King				$^3/_8$	3 3	$2^3/_4$ $2^5/_8$	$9^5/_8$ $9^5/_8$	5C = 16
Queen				$^3/_8$	3	$2^3/_4$	8	5C = 16

104F03.tif

Figure 3. Sizes Of Bricks

The modular spacing rule, *Figure 4*, does not have a gauge at the end to measure brick size. The modular scale is on one side with inches marked into sixteenths on the other.

The black figures are the references for the nominal sizes of modular brick and block. The modular markings give course numbers for different sizes. Scale 2 is for regular block, or any brick with 2 courses equal to 16 inches in height. Scale 3 measures 3 courses in 16 inches, and so on. Again, the specifications are the place to find the brick size or planned course height.

Reference number 6 equals height of each course for standard bricks.

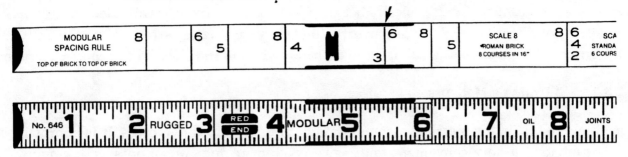

Reference number 8 equals height of each course for Roman bricks.

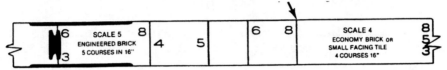

Reference number 5 equals height of each course for engineered bricks.

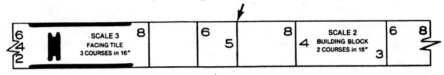

104F04.tif

Figure 4. Reading The Modular Spacing Rule

Use a modular or a course spacing rule to solve the following exercise. You will also need the information in *Figure 3*.

Exercise 9   You are building a wall 6 feet high. How many courses of brick will you need if you use:

- Modular brick
- Roman brick
- Standard building block
- Norman brick

2.3.0   DENOMINATE METRIC MEASUREMENTS

The metric system is another type of denominate measurement set. Long used worldwide, you will find this system being used more frequently in the United States. Two laws passed in the past twenty years have affected our use of the metric system. The Metric Conversion Act of 1975 called for voluntary use of the metric system. The Omnibus Trade and Competitiveness Act of 1988 declared the metric system to be the preferred system for U.S. trade. The act ordered all federal government agencies to make the change to metrics over a period of years.

Private industry, however, did not change until the General Service Administration, which oversees all federal building projects, began rejecting bids from companies that submitted nonmetric specifications. Efforts to convert to metric were further strengthened in September 1996, when all federally-assisted highway construction was required to use metric standards.

2.3.1 SI Units Of Measure

The official name of the metric system is Système International d' Unités. It is abbreviated as SI. SI metric is a very convenient and logical system of measurements and weights. It is based on the number 10, in a similar fashion to our system of currency: 10 pennies are equal to a dime, and 10 dimes are equal to a dollar.

Terms such as kilometer, centimeter, and millimeter are denominate units of metric measure. **SI metric units** of measure include base units such as meter for length, gram for weight, and liter for liquid volume. The base unit is joined with a prefix that expresses a multiple of 10. *Figure 5* below lists the more common SI prefixes.

Multiplication Factors	Prefix	Symbol
1 000 000 000 000 = 10^{12}	tera	T
1 000 000 000 = 10^{9}	giga	G
1 000 000 = 10^{6}	mega	M
1 000 = 10^{3}	kilo	k
100 = 10^{2}	hecto	h
10 = 10^{1}	deka	da
BASE UNITS 1 = 10^{0}		
0.1 = 10^{-1}	deci	d
0.01 = 10^{-2}	centi	c
0.001 = 10^{-3}	milli	m
0.000 001 = 10^{-6}	micro	μ
0.000 000 001 = 10^{-9}	nano	n
0.000 000 000 001 = 10^{-12}	pico	p
0.000 000 000 000 001 = 10^{-15}	femto	f
0.000 000 000 000 000 001 = 10^{-18}	atto	a

Figure 5. Common SI Prefixes

An SI measurement, then, is a base unit plus a prefix. The prefixes are used with all base units. A kilometer is 1,000 meters, while a millimeter is $\frac{1}{1,000}$ of a meter. A kilogram is 1,000 grams, while a milligram is $\frac{1}{1,000}$ of a gram.

2.3.2 Converting SI Metric

Until the SI metric system is adopted completely in the United States there will be a need to convert measurements from one system to the other. How do millimeters, centimeters, and meters compare with the units in the U.S. system? A centimeter is about ⅜ of an inch; a meter is about 1.1 yards; a kilometer is just a little more than 0.6 of a mile.

Figure 6 shows some common U.S.-to-metric conversions. To convert from **U.S. customary** to SI metric, follow this procedure:

- Change the U.S. measurement from fractions to decimal form.
- Multiply the decimal by the factor in the right-hand column of *Figure 6*.
- Round the answer to the accuracy required.

U.S. to Metric Conversions

Lengths
1 inch	= 2.540 centimeters
1 foot	= 30.48 centimeters
1 yard	= 91.44 centimeters or 0.9144 meters
1 mile	= 1.609 kilometers

Areas
1 square inch	= 6.452 square centimeters
1 square foot	= 920.0 square centimeters or 0.0929 square meters
1 square yard	= 0.8361 square meters

Volumes
1 cubic inch	= 16.39 cubic centimeters
1 cubic foot	= 0.02832 cubic meter
1 cubic yard	= 0.7646 cubic meter

Weights
1 ounce	= 28.35 grams
1 pound	= 435.6 grams or 0.4536 kilograms
1 (short) ton	= 907.2 kilograms

Liquid Measurements
1 (fluid) ounce	= 0.095 liter or 28.35 grams
1 pint	= 473.2 cubic centimeters
1 quart	= 0.9263 liter
1 (US) gallon	= 3785 cubic centimeters or 3.785 liters

Power Measurements
1 horsepower	= 0.7457 kilowatt

Temperature Measurements
To convert degrees Fahrenheit to degrees Celsius, use the following formula: $C = 5/9 \times (F-32)$

104F06.EPS

Figure 6. U.S. To Metric Conversions

Figure 7 shows metric-to-U.S. conversions. To convert from metric to U.S. customary, follow this procedure:

- Multiply the quantity by the factor in the right-hand column of *Figure 7*.
- Round the answer to the accuracy required.
- Convert the decimal part of the answer to the nearest common fraction.

Metric to U.S. Conversions

Lengths
1 millimeter (mm)	= 0.03937 inches
1 centimeter (cm)	= 0.3937 inches
1 meter (m)	= 3.281 feet or 1.0937 yards
1 kilometer (km)	= 0.6214 miles

Areas
1 square millimeter	= 0.00155 square inches
1 square centimeter	= 0.155 square inches
1 square meter	= 10.76 square feet or 1.196 square yards

Volumes
1 cubic centimeter	= 0.06102 cubic inches
1 cubic meter	= 35.31 cubic feet or 1.308 cubic yards

Weights
1 gram (G)	= 0.03527 ounces
1 kilogram (kg)	= 2.205 pounds
1 metric ton	= 2205 pounds

Liquid Measurements
1 cubic centimeter (cm^3)	= 0.06102 cubic inches
1 liter (1000 cm^3)	= 1.057 quarts, 2.113 pints, or 61.02 cubic inches

Power Measurements
1 kilowatt (kW)	= 1.341 horsepower (hp)

Temperature Measurements
To convert degrees Celsius to degrees Fahrenheit, use the following formula: $F = (9/5 \times C) + 32$

104F07.EPS

Figure 7. Metric To U.S. Conversions

2.3.3 Conversion Examples

Example 3   A spread footing for a foundation measures 1 foot, 6 inches wide by 22 feet long. What are its metric dimensions?

Step 1   Convert fractional units to decimal.

1 foot, 6 inches = 1⁶⁄₁₂ or 1.5 feet

Step 2   Multiply by the correct conversion factor using *Figure 6*.

1.5 feet × 30.48 centimeters =   45.7 centimeters

22 feet × 30.48 centimeters = 670.6 centimeters

Step 3   Convert SI units to the most appropriate. Centimeters are too small for measuring on a large scale. If a number is less than one, add a leading zero. Round to the nearest hundredth.

100 centimeters to a meter

45.7 centimeters ÷ 100 = 0.457 meters

670.6 centimeters ÷ 100 = 6.706 meters

The footing is 0.46 × 6.71 meters

Example 4   A drawing shows the height of a wall to be 2.5 meters. What is its height in feet and inches?

Step 1   Multiply by the correct conversion factor using *Figure 7*.

2.5 meters × 3.281 feet = 8.2025 feet

Step 2   Convert the decimal portion of feet to inches.

0.2025 feet × 12 inches = 2.43 inches = 2 inches + 0.43 inches

Step 3   Convert the 0.43 inches to a fraction.

0.43 × 16 = 6.88, round to ⁷⁄₁₆ inches

2.5 m = 8 feet, 2 ⁷⁄₁₆ inches

Use the information in *Figure 7* to complete the following exercises.

Exercise 10   Convert the following wrench sizes to their metric equivalent:

½-inch

⅝-inch

⁹⁄₁₆-inch

Exercise 11   Convert the following measurements to their metric equivalent:

13 inches

3 feet, 9 inches

your height – to centimeters and to meters

2.4.0   PLANE FIGURES AND AREA MEASURES

Knowledge of geometry is useful on a construction site. For example, if a wall is to have two windows in it, it will not need bricks in these areas. Geometry allows the mason to calculate how many less bricks will be needed. This information saves time, saves money when ordering materials, and can also be used to save steps when carrying bricks to the workstation.

The next sections review calculating areas for common geometric shapes. The measurements of these shapes may be denominated in square meters or square feet, square centimeters or square inches, depending on the job.

Plane figures are figures drawn in only two dimensions. Rectangles, triangles, and circles are common plane figures. The area of a plane figure is expressed in square units of the appropriate denomination.

2.4.1   Four-Sided Figures

Squares, rectangles, and parallelograms are four-sided regular polygons. They are figures with opposite parallel sides of the same length.

A rectangle is a polygon that has four sides of two different lengths that meet at right angles. The formula for finding the area of a rectangle is length × width, or:

$a = lw$

A square has four sides of the same length that meet at right angles. The formula for finding the area of a square is also length × width, simply expressed as side times side, or:

$a = s^2$

or

$a = ss$

A parallelogram has sides do not meet at right angles. *Figure 8* shows a parallelogram, with the "h" indicating height. The width of a parallelogram is not used in calculating area because the angles are not right angles.

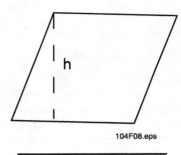

104F08.eps

Figure 8. Parallelogram

The formula for finding the area of a parallelogram is base × height. The base is the longest side, and the height is the shortest distance between the upper and lower bases. This formula is expressed as:

a = bh

2.4.2 Three-Sided Figures

Triangles are three-sided figures. The only hard rule about triangles is that the three internal angles between the sides must add up to 180 degrees. Triangles have different names according to the relationships of the sides, as shown in *Figure 9*. They also have different names according to the angles, as shown in *Figure 10*.

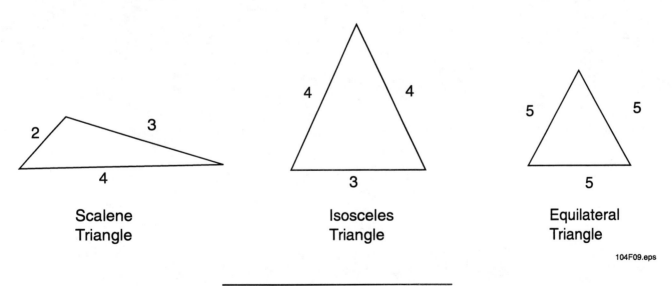

104F09.eps

Figure 9. Triangles Named By Sides

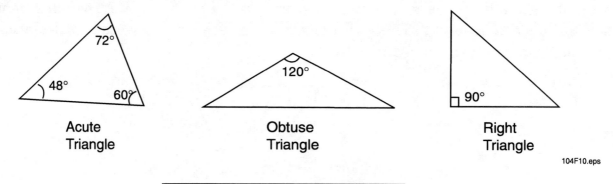

Acute
Triangle

Obtuse
Triangle

Right
Triangle

104F10.eps

Figure 10. Triangles Named By Angles

Every triangle is really an exact half of a parallelogram, as shown in *Figure 11*. Remembering this makes it easy to calculate the area of a triangle.

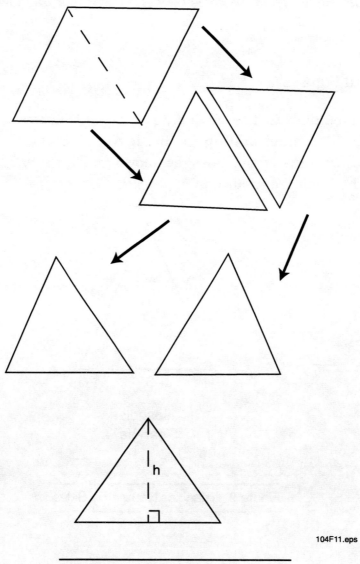

104F11.eps

Figure 11. Parallelogram To Triangle

The height of a triangle is the length of a line drawn from one angle to the side opposite, or base, that meets the base at a right angle. The area of a triangle is half the area of a parallelogram with the same base and height (*Figure 11*). The area of a triangle is expressed as:

$$a = bh/2$$

2.4.3 One-Sided Figures

Circles are single, closed lines with all points the same distance from the center. *Figure 12* shows the parts of a circle. Note that the radius is half of the diameter. Either one of these dimensions may be given on a plan or drawing to indicate the size of the circle. The circumference is the outside line that defines the circle; the area is the space within it.

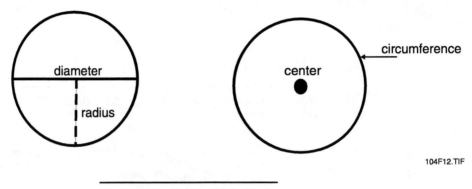

104F12.TIF

Figure 12. Parts Of A Circle

If you are building a circular garden wall, you will need to know its circumference to figure out how much material will be needed. The circumference of a circle is its diameter (d) times the constant π, or twice its radius (r) times the constant π. The rounded value of π is 3.14, or $^{22}/_{7}$. The formula for the circumference of a circle is expressed as:

$$C = \pi d$$

or

$$C = 2\pi r$$

The circumference measure is not given as squared units because the circumference measure is not an area.

You may need to know the area of a circular patio to be paved with bricks. The area of a circle is expressed in square units. The formula for the area of a circle is expressed as:

$$a = \pi r^2$$

or

$$a = \pi r r$$

2.4.4　Many-Sided Figures

Many-sided figures with all sides the same size and the same distance from the center are called regular polygons. They are also named after their number of sides. A five-sided figure is a pentagon, a six-sided figure is a hexagon, and an eight-sided figure is an octagon. *Figure 13* shows some regular polygons.

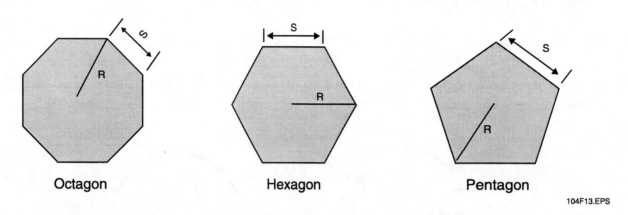

Octagon　　　　　　　　Hexagon　　　　　　　Pentagon

104F13.EPS

Figure 13.　Regular Octagon, Hexagon, And Pentagon

Occasionally, a plan will incorporate a hexagonal window over a door or an octagon in a bathroom. Sometimes a structure will be in the shape of an octagon instead of a square. To calculate the circumference of any regular polygon, you need to know the number and the length of each side.

The formula for the area of a regular polygon is the sum of the lengths of the sides divided by 2 then multiplied by r. The "r" is the distance from the center to any one angle. You may have to use your ruler on the plan to approximate that distance. The formula for the area of a regular polygon is written as:

$$A = \frac{(S_1 + S_2 + S_3 + S_n)\ r}{2}$$

The small "n" indicates that the sides continue to the required number.

Exercise 12　Calculate the areas of the following plane figures. Then convert the answers to the other system of measure.

　　a square with a height of 1.56 meters

　　a rectangle twice as wide as it is high, with a height of 3 feet, 8 inches

　　a triangle with a base of 35 centimeters and a height of 0.5 meters

　　a circle with a diameter of 0.52 meters

　　a pentagon with a side of 24 inches and r of 28½ inches

Exercise 13   You have a 3 feet wide by 6 feet high opening in a wall that was originally a window. The plans changed, and now you have to fill that space with standard block with an 8 × 16 inch face. How many blocks do you need to carry?

If you add 5 percent for breakage, how many blocks does that make?

Exercise 14   If you know that 1.125 standard block is needed to fill 1 square foot, how would you calculate *Exercise 13* differently? What would the answer be?

What if you remeasured the opening and found it to be 3 feet, 4 inches by 6 feet, 4 inches the second time you measured it? How many blocks will you carry?

Exercise 15   You need to lay a brick veneer wythe in front of that same block wall in *Exercise 13*. If you are using a modular brick with dimensions of 4 × 2⅔ × 8 inches, how many bricks will you need to carry?

If you use a Flemish bond pattern, you will need ⅓ more bricks. How many in all? How many after you remeasured the wall?

2.5.0   SOLID FIGURES AND VOLUMES

We calculate areas by measuring two-dimensional figures. The measures are usually of length and width. The calculations are written as square measurements, such as square feet or square meters. We calculate volumes by measuring three-dimensional figures. The measures are of height, width, and depth. The calculations are written as cubic measurements, such as cubic feet or cubic meters.

Figure 14 shows the most common solid figures in construction and the formulas for their volume calculations. Notice that these three figures—the rectangular solid, prism, and cylinder—have the same shape from top to bottom.

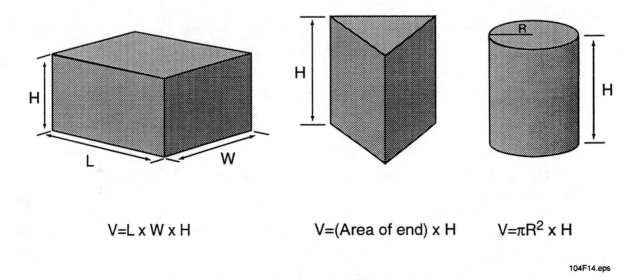

$$V = L \times W \times H \qquad V = (\text{Area of end}) \times H \qquad V = \pi R^2 \times H$$

104F14.eps

Figure 14.   Volumes Of Solid Figures

Solid figures with the same shape from top to bottom are easy to measure. Calculate the volume by first getting the area of the top surface. Then, multiply that area by the depth to get the volume.

- Volume of the rectangle = length × width × depth
    $$V = (lw)d$$

- Volume of the prism = (base × height divided by 2) × depth
    $$V = [(bh) \div 2]d$$

- Volume of the cylinder = π × the radius squared × depth
    $$V = (\pi r^2)d$$

The three figures shown in *Figure 15* are more complicated. The need to deal with the volume of these figures is rare. The formulas first calculate the area of one part of the figure, then make adjustments to the depth to get the volume.

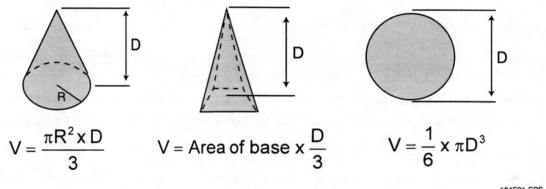

$$V = \frac{\pi R^2 \times D}{3} \qquad V = \text{Area of base} \times \frac{D}{3} \qquad V = \frac{1}{6} \times \pi D^3$$

104F21.EPS

Figure 15. Conical, Triangular, And Spherical Solids

Although the arithmetic is somewhat different, the underlying idea is the same as for the first three solid figures. These formulas are used to calculate cubic yards or cubic feet, cubic meters or cubic centimeters for these odd shapes. Calculating the cubic volumes is an important part of ordering supplies. The next two exercises provide an example of how this is done.

Exercise 16 You have just finished a cavity wall, 8 feet × 20 feet. You have a 4-inch cavity to fill with granular styrene insulation. How many cubic yards of insulation will you need?

Exercise 17 You need to reinforce a single-wythe block wall 26 feet, 8 inches long. The block is 8 × 16 inches set 9 courses high. The blocks have two cores each, and each core measures 2 × 4 × 8 inches. How many cubic yards of grout will you need to fill every core in this wall?

3.0.0 DRAWINGS

Architectural drawings are always part of project documentation. With the specification drawings, they form the written guidelines for the builder. In order to do well in your work, you must be able to read and understand the information on the project drawings. This section is a review of blueprint material in the Core Curricula. It also presents some new information about drawings, their organization, and the symbols for construction materials.

3.1.0   UNDERSTANDING DRAWINGS

In order to read drawings you have to recognize the different lines and symbols used in their preparation. The next sections review the keys to lines, symbols, and scales, and the information they carry.

3.1.1   Lines As Symbols

Each line symbol used on drawings means something different. *Figure 16* shows the most common types of lines, sometimes called the "alphabet" of lines. The bullet items that follow the figure discuss each line type.

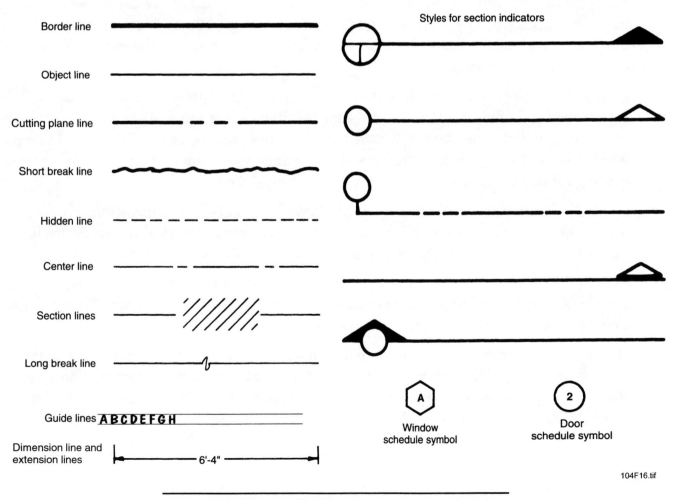

104F16.tif

Figure 16. Alphabet Of Lines For Construction Drawings

- Border lines are heavy lines used to form the boundary at the edge of the drawing. These lines define the limits of the drawings and give a finished look to the work.

- Object lines are the most common lines used on a drawing. They are used to outline the visible elements on the plan. Walls, doors, windows, roof lines, and foundations are all drawn with object lines.

- Cutting plane lines show where a cross-sectional view cuts through an elevation. Cutting plane lines are heavy lines that show where an object is cut to show internal details.

- Short break lines are heavy lines that are usually drawn freehand. They show where part of an object is peeled away to reveal a hidden feature. Examples would be part of a wall peeled to show the studs, or part of a deck peeled to show the joists.

- Hidden lines are broken lines that represent objects not visible on that particular drawing. Examples would be a foundation or a ceiling detail shown on a floor plan. One is below, while the other is above, the floor. Each needs to be shown on the floor plan to establish its relationship to other objects.

- Center lines are thin broken lines drawn through the center of symmetrical objects. Center lines make it easier to locate objects on the drawings and simplify dimensioning.

- Section lines, sometimes called crosshatch lines, are very thin lines that show an object has been sectioned. These section lines are usually drawn at a 45-degree angle, $\frac{1}{16}$ to $\frac{1}{8}$ of an inch apart. Section lines may also represent specific material, as shown in the section *Symbols and Abbreviations*.

- Long break lines are thin lines that indicate a discontinuity or disconnection. This disconnection is only on the plan, not on the item. Long break lines show items that cover longer distances than are possible given the scale of the drawing. An example would be the dimension from the edge of the foundation to the property line. Break lines are identified by an S-shaped symbol in the center of the line.

- Guide lines are also thin lines used in the preparation of the drawing itself. These lines assist the technician in laying out the structure but are not part of the structure itself. For this reason, they are usually erased and are not visible on a drawing.

- Section indication lines are sometimes called continuation lines. They are generally numbered and refer to a separate detailed drawing on another page. They provide details of that object necessary to the contractor.

- Dimension lines are thin lines used to dimension an object or to show a distance between objects. Dimension lines have three parts: a line, a dimension, and a termination symbol. Both the line and the dimension may be placed inside or outside the object, depending on the available room. *Figure 17* shows some common dimension line styles.

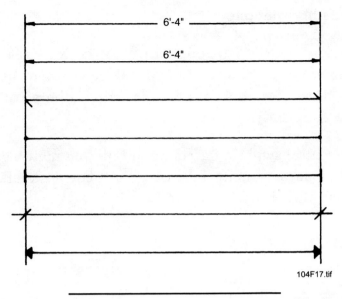

Figure 17. Dimension Line Styles

- Geometric lines are usually found in section drawings. They provide information about the shape and dimension of objects. *Figure 18* shows geometric lines usually found in drawings.

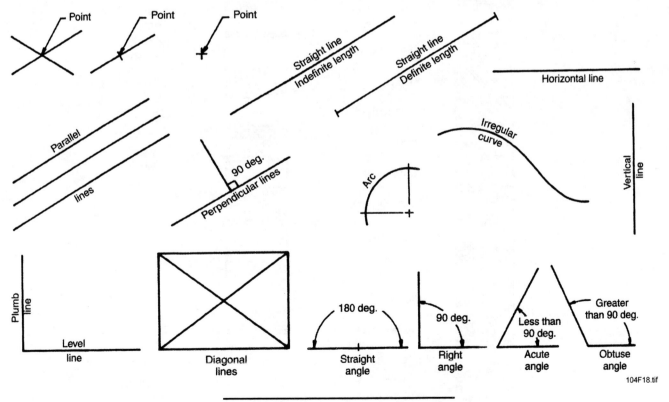

Figure 18. Geometric Drawing Lines

3.1.2 Symbols And Abbreviations

Symbols and abbreviations are used throughout a set of drawings. They convey accurate, concise, and specific information about a certain object without using much space. Some symbols will look like the object they represent while others will not. You will tend to memorize those you use frequently.

Figure 19 shows architectural symbols for common building materials. Note that about one-half the symbols refer to masonry materials. Some symbols look alike, so it is important to check where they are and what is next to them.

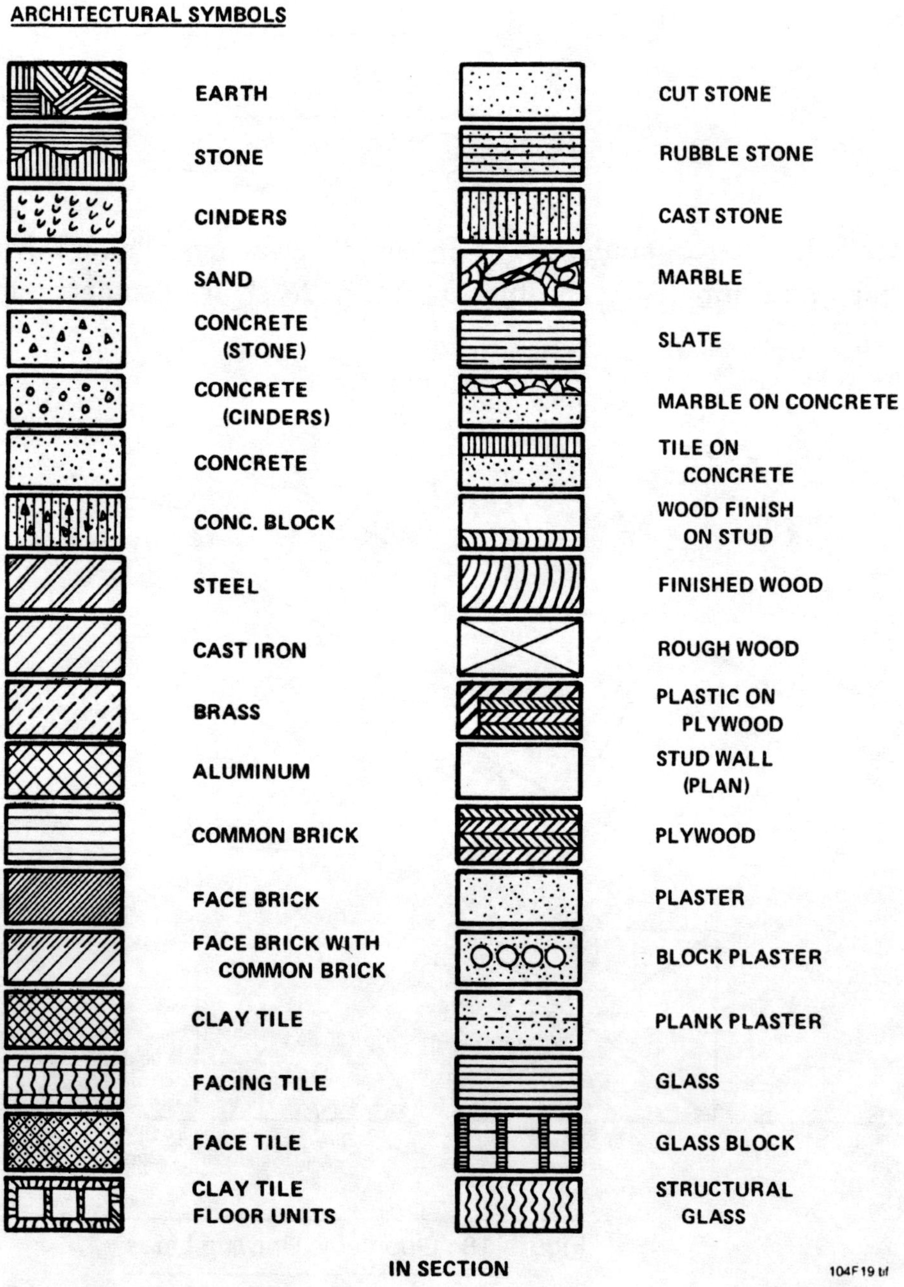

ARCHITECTURAL SYMBOLS

EARTH	CUT STONE
STONE	RUBBLE STONE
CINDERS	CAST STONE
SAND	MARBLE
CONCRETE (STONE)	SLATE
CONCRETE (CINDERS)	MARBLE ON CONCRETE
CONCRETE	TILE ON CONCRETE
CONC. BLOCK	WOOD FINISH ON STUD
STEEL	FINISHED WOOD
CAST IRON	ROUGH WOOD
BRASS	PLASTIC ON PLYWOOD
ALUMINUM	STUD WALL (PLAN)
COMMON BRICK	PLYWOOD
FACE BRICK	PLASTER
FACE BRICK WITH COMMON BRICK	BLOCK PLASTER
CLAY TILE	PLANK PLASTER
FACING TILE	GLASS
FACE TILE	GLASS BLOCK
CLAY TILE FLOOR UNITS	STRUCTURAL GLASS

IN SECTION

104F19 bf

Figure 19. Architectural Materials

Figure 20 shows electrical and plumbing symbols. Sometimes outlets or drains are located in or through masonry structures. It is important to check the drawing for any masonry structure to make sure there are no inserts for outlets or drains. If inserts are specified, they must be located where specified so they will connect properly.

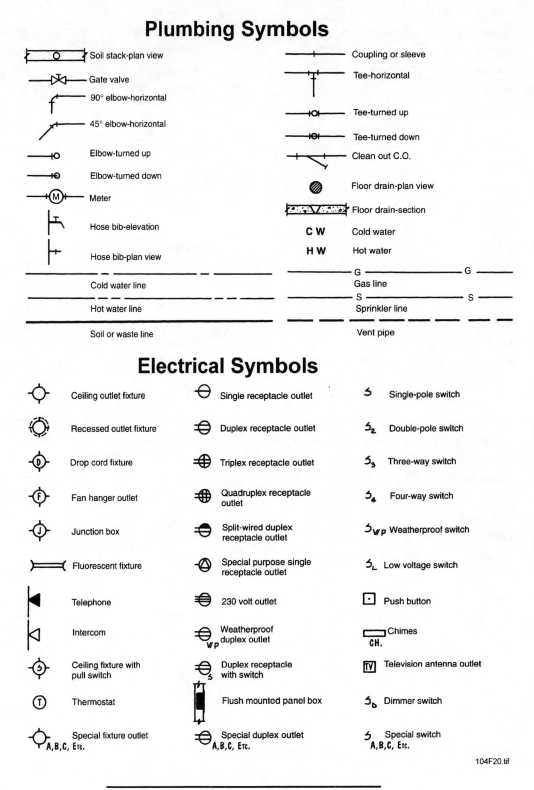

Plumbing Symbols

Soil stack-plan view	Coupling or sleeve
Gate valve	Tee-horizontal
90° elbow-horizontal	Tee-turned up
45° elbow-horizontal	Tee-turned down
Elbow-turned up	Clean out C.O.
Elbow-turned down	Floor drain-plan view
Meter	Floor drain-section
Hose bib-elevation	C W Cold water
Hose bib-plan view	H W Hot water
Cold water line	Gas line
Hot water line	Sprinkler line
Soil or waste line	Vent pipe

Electrical Symbols

Ceiling outlet fixture	Single receptacle outlet	Single-pole switch
Recessed outlet fixture	Duplex receptacle outlet	Double-pole switch
Drop cord fixture	Triplex receptacle outlet	Three-way switch
Fan hanger outlet	Quadruplex receptacle outlet	Four-way switch
Junction box	Split-wired duplex receptacle outlet	Weatherproof switch
Fluorescent fixture	Special purpose single receptacle outlet	Low voltage switch
Telephone	230 volt outlet	Push button
Intercom	Weatherproof duplex outlet	Chimes
Ceiling fixture with pull switch	Duplex receptacle with switch	Television antenna outlet
Thermostat	Flush mounted panel box	Dimmer switch
Special fixture outlet A,B,C, Etc.	Special duplex outlet A,B,C, Etc.	Special switch A,B,C, Etc.

104F20.tif

Figure 20. Electrical And Plumbing Symbols

Figure 21 is a list of common architectural abbreviations. These appear to clarify details on drawings, especially section drawings. They also appear in specification lists.

access door	AD	exterior	EXT
access panel	AP	feet	FT
acoustical tile	AT	finished floor	FIN. FL
aggregate	AGGR	firebrick	FBRK
anchor bolt	AB	fireplace	FP
angle	L	flashing	FL
approximate	APPROX	floor	FL
barrel	bbl or BBL	footing	FTG
basement	BSMT	foundation	FND
beam or bench mark	BM	glass block	GL BL
blocking	BLKG	grade	GR
blueprint	BP	height	HB
board	BD	hot water	HW
brick	BRK	I beam	I
building	BLDG	inch	in or "
building line	BL	insulation	INS
cast iron	CI	level	LEV
cast stone	CS	lineal feet	lin ft
catch basin	CB	louver	LV
caulking	CLKG	manhole	MH
ceiling height	CLG HT	marble	MR
cement mortar	CEM MORT	masonry opening	MO
centerline	C or CL	modular	MOD
center to center	C to C	not to scale	NT or NTS
channel	CHAN	on center	OC
cinder block	CB	opening	OPG
cleanout door	COD	partition	PTN
concrete	CONC	plaster	PLAS
concrete masonry unit	CMU	plate glass	PL GL
column	COL	precast	PRCST
common	COM	reinforced	REINF
contractor	CONTR	rough opening	RGH OPNG
courses	C	schedule	SCH
cross section	X-SECT	specifications	SPEC
dampproofing	DP	wall vent	WV
detail	DET	weatherproofing	WP
elevation	el or EL	weephole	WH
excavate	EXC	window	WDW
expansion joint	EXP JT		

104F21.EPS

Figure 21. Architectural Abbreviations

3.1.3 Scales And Dimensions

Drawings are normally made using a certain scale. The inches or fractions of inches on the drawing represent real distances on the project. For residential drawings, the scale is usually ¼ of an inch equals 1 foot. This means that ¼ of an inch on the drawing equals 1 foot on the ground. For larger commercial projects, a smaller scale of ⅛ of an inch to 1 foot is normally used. Architectural renderings may use a scale of ¹⁄₁₀ of an inch to 1 foot. For detail or sectional drawings, a larger scale is used. Commonly, these detail scales are ½ inch to 1 foot, or 1 inch to 1 foot. These scales are referred to as the ¼, ⅛, ½, and 1 inch scales. When reading a drawing, it is a good idea to first check the scale.

Another term you will hear is "size." The actual size of an object is two times larger than what is shown on a half-size drawing, four times larger than on a quarter-size drawing, and so on. Size drawings usually show small details of assembly or finishing. Scale and size both need to be checked when reading a drawing.

Dimensions tell the builder how large an object is and its specific location in relation to other objects. Dimensions are usually shown in feet and inches (12'-6") while dimensions less than 1 foot are shown in inches (6½"). Dimensioning interior walls is generally done in one of three ways. They are dimensioned to the center of the stud, to the outside of the stud, or to the outside of the finished walls. Dimensioning to the center of the studs is the most common.

Masonry walls are dimensioned to the outside of the walls, never to the center. *Figure 22* shows some typical dimensioning practices. Note that windows and doors in wood frame walls are dimensioned to the outside edge of the exterior wall. Windows and doors in brick veneer and frame walls are dimensioned to the edge of the framing, not the edge of the brick.

When reading a set of drawings to determine a particular dimension, do not measure the drawing. It is best to calculate that dimension from the information on the drawing. Do not measure drawings because they may have shrunk or stretched from the reproduction process. If you must measure a drawing to determine a dimension, do it in several places if possible. If you are using a reduced set of drawings (they should be marked as reduced) never measure a dimension—always calculate it.

3.2.0 RESIDENTIAL DRAWINGS

This section discusses the set of construction drawings included with this module. These working drawings are typical of those used for residential jobs. They are numbered from 1 through 11. Detailed mechanical and plumbing drawings have been intentionally omitted.

Working drawings are a set of drawings that describe a construction project in sufficient detail to allow a contractor to bid the work and then build it. The drawings also show the craftworker exactly what is to be done. Just as a mason always uses a level to check for plumb, a mason always uses a drawing to check for detail, location, and measurements.

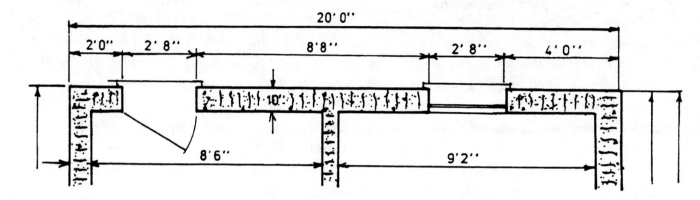

Concrete Block Walls

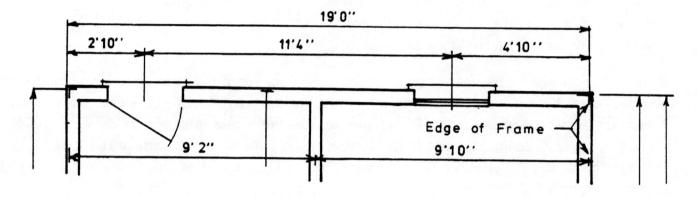

Wood Frame Walls

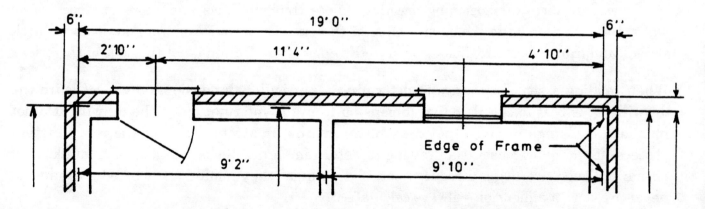

Brick Veneer/Frame Walls

104F22.TIF

Figure 22. Dimension Lines For Different Wall Types

These drawings are usually bound together in a package. The package contains a cover page, a table of contents page, and sheets showing the technical details of the project. Depending on the complexity of the project, the specifications may also be included as part of the working drawings. The title block of the drawings also contains useful information about revisions to the plans.

The number of drawings in a set of plans will vary with the complexity of the project. A private home will require fewer sheets than a commercial building, and a large industrial site will require even more drawings. A large commercial project may have site plans by the surveyor, architectural drawings by the architect, structural drawings by the structural engineer, mechanical drawings by the mechanical engineer, and electrical drawings by the electrical engineer, plus detailed specification sheets that are separate from the drawings.

Working drawings are frequently called construction drawings or blueprints. Working drawings are normally organized in the following order:

- Site or plot plans show the general layout of the land and information provided by the surveyor or site planner.
- Architectural/engineering plans include floor plans, foundation plans, elevations, section plans, roofing plans, and schedules for doors, windows, and other equipment.
- Specialty plans include mechanical, plumbing, electrical, and ductwork drawings. These may also include details for any custom features or unusual designs.

3.2.1   Plot Or Site Plan

The plot or site plan shows the location of the building in relation to the property lines. It may show utilities, contour lines, site dimensions, other buildings on the property, walks, drives, and retaining walls. This plan also shows the finished floor elevation(s) and the North direction arrow.

Plan 1 is the site plan for this project. The curved lines with numbers on the front and back lawns are elevation or contour lines. They show that the front of the lot slopes down on the right. At the left side of the house, the elevation lines show that the lot slopes very sharply down to the backyard.

3.2.2   Floor Plans

The floor plan provides the largest amount of information, perhaps making it the most important drawing of all. The floor plan shows all exterior and interior walls, doors, windows, patios, walks, decks, fireplaces, built-in cabinets, and appliances. The floor plan is actually a cross-sectional view taken horizontally between the floor and the ceiling. The height of the cross-section is usually cut about 4 feet above the floor. Sometimes this is varied to show important details of the structure.

Plans 2 and 3 are the floor plans for this project. Each level gets its own plan. Look at the detail for the walls. Note that this is the symbol for face brick, as shown previously in *Figure 19*. The doors are noted by diamonds, the windows by circles. The letters inside the symbols refer to the drawing schedules, with sizes and other details listed in table form.

Looking at the plans, how many windows and doors are scheduled for level 1? For level 2? Is the fireplace to be of brick on the inside? Can you name some of the types of lines used on this plan? How many different lines are used on this plan?

3.2.3 Foundation Drawings

The foundation plan is usually the first of the structural drawings. This shows the foundation size and materials. It shows details about excavation, waterproofing, and supporting structures such as footings or piles. It can have sections of the footings and foundation walls. When a building has a basement, it is included here as well.

Plan 4 is the foundation plan. Notice that it has two section indicators, C/C and D/D. Section indicator D/D crosses a line for a hidden feature, a thickened area in the slab. The section drawings called out by those indicators are in the upper right corner of the plan. In the section detail, what is the distance set for the weepholes? What does O.C. mean? What do the dashed squares in the upper middle mean?

3.2.4 Elevation Drawings

The elevation drawings are views of the exterior features of the building. They usually show all four sides of a building. Sometimes, for a building of unusual design, more than four elevations may be needed. The elevation drawings show outside features such as placement and height of windows, doors, chimneys, and roof lines. Exterior materials are indicated as well as important vertical dimensions.

Interior elevation drawings show in greater detail the various cabinets, bookshelves, fireplaces, and other important interior features. These are sometimes called detail drawings.

Plans 5 and 6 are the exterior elevation drawings. Notice in these plans that the concrete block retaining wall is now a very obvious feature of the building. Can you find it, less apparent, on the site plan? Plan 7 is an interior elevation of the kitchen, vanities, and fireplaces. What are the hidden lines showing for the fireplace elevations?

3.2.5 Section Drawings

Sections give information about the construction of walls, stairs, or other items which cannot be easily given on the elevation, floor plan, or framing drawings. These drawings are usually drawn to a scale large enough to show the details without cluttering the drawing. A section taken through the narrow width of a building is known as a transverse section. A section taken through the entire length is known as a longitudinal section.

Go back to Plan 3. Locate the section indicators A/A and B/B. Then look at Plan 8. These are the section plans for the horizontal cuts along the section indicators. The section drawings show the building from the foundation footings to the roof. Notice that the large crawl space is now a very obvious feature of the building.

Plan 9 shows sections through exterior and interior walls. This plan is at a different scale and a much finer level of detail. The working mason should be able to read this section plan and answer questions about the work. What is the foundation wall made of? What is the distance between the brick veneer and the backing walls? How frequently is the veneer tied to the backing? What is the distance between weepholes? Is there a specification here for the windowsills?

3.2.6 Schedules

Schedules are an important tool in a set of plans. Although they are tables, not drawings, they give specific information that would be impractical to include on the documents. They also make it convenient for ordering material by collecting this detailed information in one place.

Go back to Plan 7, which has the door and window schedules. What type of window will be in the kitchen? How many different sizes of door will be needed? What other kinds of items might be listed on schedules?

3.2.7 Specialty Plans

Specialty plans give details about plumbing, electrical, and duct work. It might appear that a mason does not have to know about these things, but this appearance is deceiving.

Plans 10 and 11 are the electrical plans for the first and second floors. Each floor has at least one electrical connector box required on the outside of that brick veneer wall. How many outside connector boxes are needed for both floors? What will be attached to those connector boxes? How will this influence the location of those boxes?

Masons are only one of the many kinds of craftworkers on the job. Understanding what other workers are doing, and how they affect your work, is important. This information is on the drawings. Use them to help you work smarter.

3.2.8 Metric Drawings

As discussed earlier, the metric system of measurement will eventually be the world standard. Even though no standard sizings have been set for the lumber and building industries in the U.S., some work is going on. New standard sizes for lumber and other materials must be approved. Clay and concrete masonry products have made a soft conversion. This means that the size of the products has not changed, but metric sizes have been identified. Units sized to metric dimensions will be adopted in the future. *Figure 23* shows a construction plan labeled using the metric system.

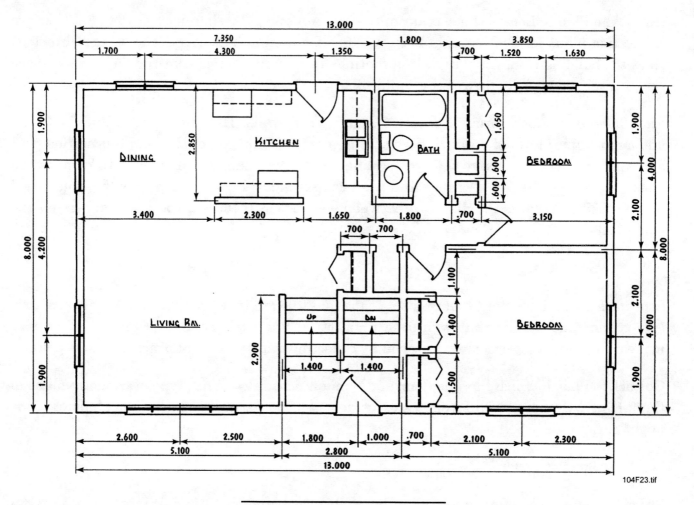

104F23.tif

Figure 23. Example Metric Plan

On this plan, units for linear measure are restricted to the meter (m) and the millimeter (mm). Thus, the figures to the left of the decimal indicate meters while the figures to the right indicate millimeters. Note that the lines and symbols used in the remainder of the plan are not changed.

4.0.0   SPECIFICATIONS, STANDARDS, AND CODES

Construction plans or drawings usually include a set of written specifications, or "specs" as they are referred to. These detail information not found on the drawings regarding the quality or minimum standards that must be met. These specifications are the written or printed instructions or information needed to complete the construction. In fact, such specifications are part of the building contract. They represent information from the architect, the engineer, and the owner of the structure.

Because the plans cannot give or show all of the information necessary for the mason to complete the work, specifications add the details. For example, specifications include information on:

- Quality of materials
- Quality of workmanship (minimum tolerance)
- Procedures or techniques to be used during construction
- Various responsibilities of each subcontractor

Generally, specifications are divided into two major areas: general conditions and technical specifications. The general conditions cover such legal items as insurance, permit responsibilities, and payment schedules. The mason will have little contact with these documents. Masons will need to read the technical specifications, though, to do their work properly.

4.1.0   TECHNICAL SPECIFICATIONS

Technical specifications deal with the actual conduct of the project and the craftwork involved. Each section includes information for the work of a single subcontractor. Technical specifications directly affect the mason's work. Masonry work has its own set of technical specifications on every project.

Because they are part of a contract, technical specifications are legal documents also. If there is a discrepancy between the drawings and the specifications, the specifications have priority. The mason must have both the plans and the specifications to construct the project. When conflicts occur between the drawings and specifications, consult the masonry supervisor or superintendent.

Figure 24 is the first page of a 4-page general technical specification form with fill-in blanks. It is used by the Federal Housing Administration (FHA) and the Veteran's Administration (VA) when they finance home building. Other agencies or private companies may use some version of this form. It specifies materials, lumber grades, framing dimensions, sills, and so on. It has a very short section on masonry and does not specify many items. Projects using this specification will supplement it with additional forms and specifications.

FHA Form 2005
VA Form 26-1852
Rev. 3/68

For accurate register of carbon copies, form
may be separated along above fold. Staple
completed sheets together in original order.

Form approved.
Budget Bureau No. 63-R055.11.

☐ Proposed Construction

DESCRIPTION OF MATERIALS

No. _____
(To be inserted by FHA or VA)

☐ Under Construction

Property address _____ City _____ State _____

Mortgagor or Sponsor _____ _____
(Name)   (Address)

Contractor or Builder _____ _____
(Name)   (Address)

INSTRUCTIONS

1. For additional information on how this form is to be submitted, number of copies, etc., see the instructions applicable to the FHA Application for Mortgage Insurance or VA Request for Determination of Reasonable Value, as the case may be.
2. Describe all materials and equipment to be used, whether or not shown on the drawings, by marking an X in each appropriate check-box and entering the information called for in each space. If space is inadequate, enter "See misc." and describe under item 27 or on an attached sheet.
3. Work not specifically described or shown will not be considered unless

required, then the minimum acceptable will be assumed. Work exceeding minimum requirements cannot be considered unless specifically described.
4. Include no alternates, "or equal" phrases, or contradictory items. (Consideration of a request for acceptance of substitute materials or equipment is not thereby precluded.)
5. Include signatures required at the end of this form.
6. The construction shall be completed in compliance with the related drawings and specifications, as amended during processing. The specifications include this Description of Materials and the applicable Minimum Construction Requirements.

1. EXCAVATION:
Bearing soil, type _____

2. FOUNDATIONS:
Footings: concrete mix _____; strength psi _____ Reinforcing _____
Foundation wall: material _____ Reinforcing _____
Interior foundation wall: material _____ Party foundation wall _____
Columns: material and sizes _____ Piers: material and reinforcing _____
Girders: material and sizes _____ Sills: material _____
Basement entrance areaway _____ Window areaways _____
Waterproofing _____ Footing drains _____
Termite protection _____
Basementless space: ground cover _____; insulation _____; foundation vents_____
Special foundations _____
Additional information: _____

3. CHIMNEYS:
Material _____ Prefabricated (make and size) _____
Flue lining: material _____ Heater flue size _____ Fireplace flue size _____
Vents (material and size): gas or oil heater _____; water heater _____
Additional information: _____

4. FIREPLACES:
Type: ☐ solid fuel; ☐ gas-burning; ☐ circulator (make and size) _____ Ash dump and clean-out _____
Fireplace: facing _____; lining _____; hearth _____; mantle _____
Additional information: _____

5. EXTERIOR WALLS:
Wood frame: wood grade, and species _____ ☐ Corner bracing. Building paper or felt _____
    Sheathing _____; thickness _____; width _____; ☐ solid; ☐ spaced _____" o. c.; ☐ diagonal; _____
    Siding _____; grade _____; type _____; size _____; exposure _____"; fastening _____
    Shingles _____; grade _____; type _____; size _____; exposure _____"; fastening _____
    Stucco _____; thickness _____"; Lath _____; weight _____ lb.
    Masonry veneer _____ Sills _____ Lintels _____ Base flashing _____
Masonry: ☐ solid ☐ faced ☐ stuccoed; total wall thickness _____"; facing thickness _____"; facing material _____
    Backup material _____; thickness _____"; bonding _____
    Door sills _____ Window sills _____ Lintels _____ Base flashing _____
    Interior surfaces: dampproofing, _____ coats of _____; furring _____
Additional information: _____
Exterior painting: material _____; number of coats _____
Gable wall construction: ☐ same as main walls; ☐ other construction _____

6. FLOOR FRAMING:
Joists: wood, grade, and species _____; other _____; bridging _____; anchors _____
Concrete slab: ☐ basement floor; ☐ first floor; ☐ ground supported; ☐ self-supporting; mix _____; thickness _____";
    reinforcing _____; insulation _____; membrane _____
Fill under slab: material _____; thickness _____". Additional information: _____

7. SUBFLOORING: (Describe underflooring for special floors under item 21.)
Material: grade and species _____; size _____; type _____
Laid: ☐ first floor; ☐ second floor; ☐ attic _____ sq. ft.; ☐ diagonal; ☐ right angles. Additional information: _____

8. FINISH FLOORING: (Wood only. Describe other finish flooring under item 21.)

LOCATION	ROOMS	GRADE	SPECIES	THICKNESS	WIDTH	BLDG. PAPER	FINISH
First floor							
Second floor							
Attic floor _____ sq. ft.							

Additional information: _____

FHA Form 2005
VA Form 26-1852

1

DESCRIPTION OF MATERIALS

109F24.TIF

Figure 24. Standard FHA/VA Specification Form

Regardless of the forms used, masonry trade specifications should include the information outlined in *Figure 25*. This information defines the mason's responsibility on the project.

Scope of the work

Type of work and materials to be used

The mason's responsibilities (for instance, is the mason responsible for applying flashings? caulking? for job cleanup?)

Type of materials

Type of face and common brick, concrete block, and stone, and type of mortar mix specified for each

Type of waterproofing for foundation wall and where it is to be applied to the wall

Quality standards which the materials should meet, such as compressive strength and color range

Masonry workmanship

Quality of workmanship expected

Type of mortar joint finish and the type of bond desired

Size of pieces of masonry units to be installed in the wall; smallest allowable piece (for example, no piece smaller than 2 inches in face brickwork). Many masonry specifications stipulate that no bats are to be used in place of headers.

Cutting instrument to be used on masonry units, such as a chisel and hammer or masonry saw

Care of the wall at the conclusion of the day's work, such as the stipulation that all masonry work be covered with tarpaulins or heavy plastic to protect it from the weather

Minimum temperature that work may commence to prevent the freezing of mortar

Specific instructions on the parging of the basement wall

Stone work

Type of stone and bond pattern desired

Size and type of mortar joint required

Method in which the stone is to be cut, chiseled, or sawed with the masonry saw

Type of protection for the work

Concrete masonry unit

Specific shape, such as bullnose or corner block

Type of mortar and mix

Type of joint finish and allowable thickness

Type of wall ties or joint reinforcement

Type of lintel to be placed over openings, such as concrete masonry lintel, steel, or precast concrete

Building in and around mechanical work

Methods of installation of pipe chases, heating units, and electrical work

Caulking

Type of caulking and filler materials to be installed in back of the caulking if necessary

Workmanship to be neat and excess caulking removed from frames

Cleaning and pointing masonry work

Type of cleaning agent to be used; instructions on how concrete block and stone are to be cleaned

Directions indicating that all holes in mortar joints are to be pointed with fresh mortar as the work is cleaned

104F25.EPS

Figure 25. Specification Contents

Highly detailed technical specifications spell out not only the materials but how they are to be assembled and finished. Doing so defines the long-term performance of the materials. This goes back to engineering studies about the properties of masonry construction.

Specifications take advantage of these engineering studies. They do this by referencing existing standards and codes. By referencing them, the specifications incorporate the provisions of the standards and codes.

Almost every type of building material has a multitude of size and performance standards and codes for it. The standards and codes guide and regulate manufacturers, designers, and builders. There are voluntary standards, national standards, international standards, national codes, local building codes, and model codes. The next sections discuss standards and codes in greater detail.

4.2.0   STANDARDS

Specifications are prepared for specific projects. They reflect the needs of that project and the wishes of the owners and the designer. They also use and incorporate local and national standards publications.

When you begin studying technical specifications, you will find items like "work to meet ASTM C144-96" or "as defined by CSA A165.2." These phrases refer to studies by the American Society for Testing and Materials (ASTM) or the Canadian Standards Association (CSA). These studies have established minimum standards for all aspects of masonry units and masonry work. A standard usually has a single subject. It defines one or more aspects of its subject. The standards are based on studies, research, and advances in materials and construction techniques. Typical standards include ASTM C1072, *Method for Measurement of Masonry Flexural Bond Strength*, and ASTM C315, *Specification for Clay Flue Linings*.

The ASTM and CSA are only two of the organizations publishing standards for the masonry industry. Consensus standards and codes are developed by other independent organizations such as the American Society of Civil Engineers (**ASCE**), the American Institute of Steel Construction (**AISC**), the National Concrete Masonry Association (**NCMA**), and the American Concrete Institute (**ACI**). You may see references made to these other standards in your specifications.

ASTM and the other organizations gather data on new materials and techniques and update their standards. For ASTM publications, the year of publication follows a dash after the title number. Appendix B lists ASTM standards for masonry construction. By reading through the list of titles in Appendix B, you can get a good idea of how much study has gone into modern masonry materials and practices.

Standards are referenced by specifications and by codes.

4.3.0 CODES

Model codes are technical documents written by the members of three code organizations. The codes are concerned with more than one subject, and more than one aspect of each subject. They are based on, or incorporate by reference, many standards published by ASTM and other standards organizations. The codes also reference other codes published by industry associations. The model codes and standards documents are available for adoption by state and local jurisdictions.

If a jurisdiction adopts a model standard, it becomes a local, not a model, standard. In addition to the model contents, local codes also include local requirements, such as conformance with a zoning plan, or lot setbacks. Most local building codes in the United States are based on one of the following model codes:

- *The Southern Standard Building Code*, published by the Southern Building Code Congress (SBCC), is typically used throughout the Southeast.

- *The National Building Code*, published by the Building Officials and Code Administrators International (**BOCA**), is adopted mostly in the northeast and central states.

- *The Uniform Building Code*, published by the International Conference of Building Officials (**ICBO**), is used throughout the West.

The map in *Figure 26* shows the areas of the United States where these codes are used.

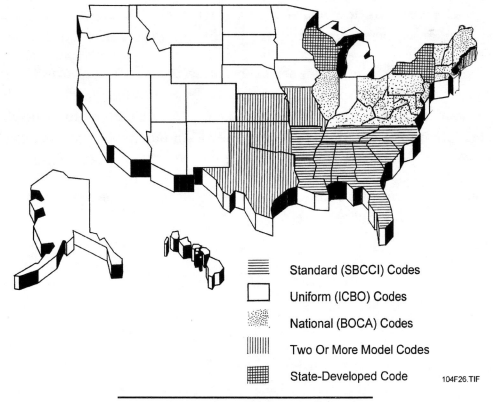

Figure 26. Distribution Of Building Codes

These codes are used in different regions because they have different areas of concern. Western builders do not need to build against hurricane damage, but they must consider earthquakes. Southern builders must deal with higher humidity, but earthquakes are not as common as hurricanes. Local or regionally focused building codes take account of these differences in their contents.

Codes and specifications are highly formal and regulated subjects. In an effort to coordinate the works of the three code organizations, a fourth organization called the Council of American Building Officials (**CABO**) was formed in 1972.

Their first task, done at the request of the home-building industry, was to develop a simple stand-alone code document that describes the most commonly accepted practices in home building. The result was the *CABO One and Two Family Dwelling Code*. This set of codes comes in a ten volume set and covers a comprehensive number of topics.

The ACI and the ASCE agreed to jointly develop a new consensus standard for masonry design with the support of the masonry industry. Public review began in 1988, and the code was approved in 1989. The Standard Building Code Congress International (SBCCI) adopted this as part of the new Standard Building Code in 1988. BOCA adopted it in 1989. The new standards replaced the previously existing masonry design and construction standards. They contribute to consistent design and a higher quality of construction. They incorporate the latest research on material properties and performance.

The ASTM and the ICBO jointly publish the *Uniform Building Code*. This code references ASTM and **UBC** standards for clay and cement masonry units, stone, mortar and grout, reinforcement and accessories, sampling and testing, and prefabricated assemblages.

As of 1997, a new standard has been set for building codes, the *Building Code Requirements for Masonry Structures*, ACI 530-92/ASCE 5-92/TMS 402-92. Jointly published by these organizations and the National Concrete Masonry Association, this consolidates several masonry codes into one consistent set of provisions. An outline of this document is presented as *Figure 27*.

Reading through the contents, remember that each of these topics refers to many standards and many years of testing and research.

Building Code Requirements for Masonry Structures

PART 1–GENERAL

Chapter 1–General Requirements

1.1 Scope
1.2 Permits and drawings
1.3 Approval of special systems of design or construction
1.4 Standards cited in this code

Chapter 2–Notation and Definitions

2.1 Notation
2.2 Definitions

PART 2–QUALITY ASSURANCE AND CONSTRUCTION REQUIREMENTS

Chapter 3–General

3.1 Materials, labor, and construction
3.2 Acceptance relative to strength requirements

Chapter 4–Embedded Items–Attachments to framing and existing construction

4.1 Embedded pipes and conduits
4.2 Attachment of masonry to structural frames and other construction
4.3 Connectors

Chapter 5–General Analysis and Design Requirements

5.1 Scope
5.2 Loading
5.3 Load combinations
5.4 Design strength
5.5 Material properties
5.6 Deflection of beams and lintels
5.7 Lateral load distribution
5.8 Multiwythe walls
5.9 Columns
5.10 Pilasters
5.11 Load Transfer
5.12 Concentrated loads
5.13 Section properties
5.14 Anchor bolts solidly grouted in masonry
5.15 Framed construction
5.16 Stack bond masonry

Chapter 6–Unreinforced Masonry

6.1 Scope
6.2 Stresses in reinforcement
6.3 Axial compression and flexure
6.4 Axial tension
6.5 Shear

Chapter 7–Reinforced Masonry

7.1 Scope
7.2 Steel reinforcement
7.3 Axial compression and flexure
7.4 Axial tension
7.5 Shear

Chapter 8–Details of Reinforcement

8.1 Scope
8.2 Size of reinforcement
8.3 Placement limits for reinforcement
8.4 Protection for reinforcement
8.5 Development of reinforcement embedded in grout

Chapter 9–Empirical Design of Masonry

9.1 Scope
9.2 Height
9.3 Lateral stability
9.4 Compressive stress requirements
9.5 Lateral support
9.6 Thickness of masonry
9.7 Bond
9.8 Anchorage
9.9 Miscellaneous requirements

Appendix A–Special Provisions for Seismic Design

A.1 General requirements
A.2 Special provisions for Seismic Zones 0 and 1
A.3 Special provisions for Seismic Zone 2
A.4 Special provisions for Seismic Zones 3 and 4

104F26.eps

Figure 27. Building Code Requirements

Specifications for commercial and industrial projects are based on one of the three national model codes, plus local codes. Unlike residential projects, commercial and industrial projects are always designed and managed by architects and engineers registered by the state. They are responsible for insuring that the project is built to plans and specifications and complies with all local code requirements. Local codes for commercial and industrial projects usually are more specific about fire protection, load bearing, and parking spaces than codes for residential buildings.

4.4.0 INSPECTION AND TESTING

Every construction contract has specifications to cover inspection and acceptance of the finished work. ASTM has published many specifications about testing. Acceptance of the finished product depends on the inspection and test results.

For residential and commercial projects, inspection and testing are usually done by the county building inspector. This inspector will be looking for compliance with the local building code. The inspector works closely with the contractor to coordinate inspections with the completion of critical phases of the work. Work cannot progress on the next phase until this inspection has been made and the inspector signs off on the work. For this reason, you may see an inspector's check-off sheet posted on the job site with the inspector's comments and signature for each phase. This may be in the job superintendent's office or it may be posted at the next place the inspector is due to visit.

For residential projects, testing is usually limited to looking at the material tags and manufacturer's documents that come with the materials. For this reason, you should not throw these documents away until the inspector visits the site and signs off on the work.

On some projects, the architect will also inspect the work as it progresses. The architect will be looking for compliance with the drawings and specifications and will not worry about the building code. On some projects, a representative of the owner may check the craftsmanship of the work accomplished every day. Work that does not meet standards, drawings, specifications, or local codes must be done again. This can be very costly. This is why it is important to do it right the first time.

Inspection and testing on commercial and industrial projects is much more involved and, depending on the project size, will require a full-time staff hired by the owner. The representatives of the owner and/or architect will look for quality and compliance with the plan and specifications. The building inspector will look for compliance with the code. The contractor must pay to take out and replace work that is not up to specification or code, just as with residential projects.

Wise masons remember this when they start a project. This is why they look at the drawings and specifications before they lift a single brick.

SUMMARY

This training module covers three topics: math, drawings, and specifications.

Math skills are useful for calculating materials and supplies, as well as in the interpretation of project drawings. Masons need to know how to convert denominate fractions and how to convert metric measures. They must be able to calculate areas and volumes of common geometric figures. They also need to read mason's rules, both modular and course.

Project drawings include several categories. A project drawing package will include site plans, floor plans, elevations, sections, materials schedules, structural drawings, and mechanical drawings. They carry information in detail by using lines and symbols, as well as measurement notations.

Project drawings also include specifications which provide information not visible on the plans. Technical specifications refer to and include standards and codes. Standards are single-topic, nationally published, detailed measures. Model codes include references to many standards. Local codes are usually based on model codes and include local matters such as zoning.

References

For advanced study of topics covered in this task module, the following books are suggested:

Building Block Walls—A Basic Guide, National Concrete Masonry Association, Herndon, VA, 1988.

Masonry Design and Detailing—For Architects, Engineers and Contractors, Fourth Edition, Christine Beall, McGraw-Hill, New York, NY, 1997.

Masonry Skills, Third Edition, Richard T. Kreh, Delmar Publishers, Inc., Albany, NY, 1990.

The ABC's of Concrete Masonry Construction, Videotape, 13:34 minutes, Portland Cement Association, Skokie, IL, 1980.

ACKNOWLEDGMENTS

Figures 1, 3, 5-15, 17, 21, and 25-27 courtesy of Roy Jorgensen Associates, Inc.

Figures 2, 4, and 22 courtesy of Associated General Contractors

Figures 16, 18, and 23 courtesy of The Goodheart-Willcox Company, Inc.

Figures 19 and 20 courtesy of Delmar Publishers

Figure 24 courtesy of Federal Housing Administration

REVIEW/PRACTICE QUESTIONS

1. Add the following:

	3 yards	1 foot	11 inches
+	1 yard	2 feet	8 inches

   a. 4 yards   3 feet   7 inches
   b. 1 yard   2 feet   3 inches
   c. 5 yards   1 foot   7 inches
   d. 4 yards   3 feet   19 inches

2. Subtract the following:

	3 gallons	2 quarts	1 pint
−	1 gallon	3 quarts	1 pint

   a. 4 gallons   5 quarts   2 pints
   b. 1 gallon   3 quarts
   c. 2 gallons   3 quarts
   d. 5 gallons   2 quarts

3. Look at the tape measure. At what point is the arrow?

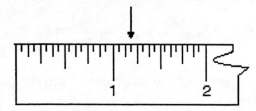

   a. $1\frac{3}{32}$ inch
   b. $2\frac{13}{16}$ inches
   c. $1\frac{3}{8}$ inch
   d. $1\frac{3}{16}$ inch

4. Measure this line.

   a. $4\frac{5}{8}$ inches
   b. $4\frac{25}{32}$ inches
   c. $4\frac{15}{16}$ inches
   d. $4\frac{2}{3}$ inches

5. Convert 2⅜ inches into the SI metric system.

   a. 2.375 centimeters
   b. 6.033 centimeters
   c. 23.75 meters
   d. 6.033 meters

6. Convert 5.378 meters to U.S. customary units (round to nearest ¹⁄₁₆ of an inch).

   a. 5 yards   0 feet   10 ⁹⁄₁₆ inches
   b. 5 yards   2 feet   7¾ inches
   c. 3 yards   2 feet   1½ inches
   d. 5 yards   1 foot   8¾ inches

7. 32.54 square feet equals _____ square meters.

   a. 2.875
   b. 3.254
   c. 5.043
   d. 3.023

8. A foundation measures 42 feet long and 12 feet wide. What is the area in square feet?

   a. 504 square feet
   b. 520 square feet
   c. 54 square feet
   d. 424 square feet

9. A wall measures 4 feet high and 40 feet long. You are going to face the wall using a brick with a nominal size of 4 inches × 2⅔ inches × 8 inches. From a chart you determine you will need 675 bricks per 100 square feet. How many bricks will be needed assuming no wastage or breakage?

   a. 2,160
   b. 972
   c. 1,080
   d. 1,180

10. A foundation is to be circular and 8 inches thick. Its diameter is 8 feet, 6 inches. How many cubic feet of concrete will it take? Round your answer to one decimal place.

    a. 151.2 cubic feet
    b. 38.7 cubic feet
    c. 39.3 cubic feet
    d. 37.8 cubic feet

11. Calculate the volume of a cone whose height is 18 inches and diameter is 12 inches. Round the answer to no decimal place.

    a. 619 cubic inches
    b. 678 cubic inches
    c. 2,034 cubic inches
    d. 2,713 cubic inches

12. Heavy lines used to form a boundary on a drawing are called _____.

    a. extension lines
    b. guide lines
    c. border lines
    d. section lines

13. Center lines are _____ and are used with _____.

    a. very fine, lettering
    b. thin broken lines, symmetrical objects
    c. thin, invisible objects
    d. thin, dimensions

14. The abbreviation INS on a drawing means _____.

    a. insert here
    b. inches
    c. insulation
    d. in shifts

15. After looking at a ¼ scale floor plan, you decide there is no way to calculate a certain dimension that you need. Using a ruler, you measure the dimension as 4⅞ inches. What dimension will you use when laying out the actual work?

    a. 19 feet, 6 inches
    b. 4 feet, 10½ inches
    c. 9 feet, 9 inches
    d. 19 feet, 3 inches

16. If you wanted details on the type of tile to be used in the bathroom, you would look on which drawings?

    a. Section drawings
    b. Elevation drawings
    c. Foundation drawings
    d. Schedule drawings

17. Ceiling heights would be shown on which drawings?

    a.   Section drawings
    b.   Elevation drawings
    c.   Foundation drawings
    d.   Schedule drawings

18. A walled patio has been added to the project but does not appear on the drawing. The walls are supposed to run north to south and east to west. Which drawing would provide you with guidelines on how to lay out the walls?

    a.   Section plan
    b.   Floor plan
    c.   Elevation plan
    d.   Plot or site plan

19. Which drawing will show you the distances from an interior wall to the foundation wall?

    a.   Floor plan
    b.   Section plan
    c.   Foundation plan
    d.   Elevation plan

20. A shopping center project in New Mexico will be governed by which set of model codes?

    a.   Standard (SBCCI) codes
    b.   Uniform (ICBO) codes
    c.   National (BOCA) codes
    d.   CABO codes

21. If a conflict is discovered between the drawings and the specifications, how is it resolved?

    a.   The drawings take priority
    b.   The specifications take priority
    c.   The building inspector must decide
    d.   The job superintendent must decide

22. A mason wants to know what the specifications say about embedded items. Which section of the *Building Code Requirements for Masonry Structures* should the mason read?

    a.   4.2
    b.   7.2
    c.   8.3
    d.   5.12

23. Add the following:

	1 yard	2 feet	8 inches
		2 feet	11 inches
+		1 foot	6 inches

   a.   2 yards   6 feet   1 inch
   b.   3 yards   0 feet   1 inch
   c.   1 yard   5 feet   25 inches
   d.   3 yards   1 foot   1 inch

24. You have a ½-inch open end wrench. What is the closest metric size? (A metric wrench size is its millimeter measurement.)

   a.   Size 10 metric
   b.   Size 11 metric
   c.   Size 12 metric
   d.   Size 13 metric

25. If a tape measure is marked in ¹⁄₁₆-inch increments, which measurement cannot be determined using the tape?

   a.   2½ inches
   b.   3⅜ inches
   c.   4⅕ inches
   d.   11¾ inches

ANSWERS TO REVIEW/PRACTICE QUESTIONS

<u>Answer</u>		<u>Section Reference</u>
1.	c	2.1.1
2.	b	2.1.2
3.	d	2.2.1
4.	b	2.2.1
5.	b	2.3.2
6.	b	2.3.2
7.	d	2.3.2
8.	a	2.4.1
9.	c	2.4.1
10.	d	2.5.0
11.	b	2.5.0
12.	c	3.1.1
13.	b	3.1.1
14.	c	3.1.2/Figure 21
15.	a	3.1.3
16.	d	3.2.6
17.	a	3.2.5
18.	d	3.2.1
19.	a	3.2.2
20.	b	4.3.0
21.	b	4.1.0
22.	a	4.3.0/Figure 27
23.	d	2.1.1
24.	d	2.3.2
25.	c	2.2.1

1. 6 yards 1 foot 3 inches

2. 11 yards 2 feet 1 inch

3. 12 yards 1 foot 1 inch

4. 13 yards 0 feet 4 inches

5. 4 yards 0 feet 10 inches

6. 1 yard 2 feet 10 inches

7. 3 yards 0 feet 10 inches

8. 3 yards 0 feet 9 inches

9. | | |
    |---|---|
    | Modular brick | 27 courses |
    | Roman brick | 36 courses |
    | Standard building block | 9 courses |
    | Norman brick | 27 courses |

10. ½ inch = 12.70 mm

    ⅝ inch = 15.88 mm

    9⁄16 inch = 14.29 mm

11. 13 inches = 33.02 cm

    3 feet, 9 inches = 1 meter, 14.3 cm

    Have another trainee check your height

12. | | |
    |---|---|
    | Square | 2.4336 square meters; 26.19 square feet |
    | Rectangle | 26.8889 square feet; 2.498 square meters |
    | Triangle | 875.0 square cm; 135.63 square inches |
    | Circle | 0.2123 square meters; 2.284 square feet |
    | Pentagon | 1,710 square inches or 11.875 square feet |
    | | 11,032.92 square centimeters or 1.1032 square meters |

13. 20.25 blocks ≅ 21 blocks

    21.26 blocks ≅ 22 blocks

14. 20.25 ≅ 21 blocks

    24 blocks

15. | | |
    |---|---|
    | Modular brick | 121.5 ≅ 122 bricks |
    | Flemish bond pattern | 162 bricks |
    | Remeasure modular brick | 145 bricks |
    | Remeasure Flemish bond | 184 bricks |

16. 1.9753 cubic yards

17. 0.4938 = 0.5 cubic yards

Clay Masonry Units

ASTM C27, *Fire Clay and High Alumina Refractory Brick*

ASTM C32, *Sewer and Manhole Brick*

ASTM C34, *Structural Clay Loadbearing Wall Tile*

ASTM C43, *Terminology Relating to Structural Clay Products*

ASTM C56, *Structural Clay Non-Loadbearing Tile*

ASTM C62, *Building Brick*

ASTM C106, *Fire Brick Flue Lining for Refractories and Incinerators*

ASTM C126, *Ceramic Glazed Structural Clay Facing Tile, Facing Brick and Solid Masonry Units*

ASTM C155, *Insulating Fire Brick*

ASTM C212, *Structural Clay Facing Tile*

ASTM C216, *Facing Brick*

ASTM C279, *Chemical Resistant Brick*

ASTM C315, *Clay Flue Linings*

ASTM C410, *Industrial Floor Brick*

ASTM C416, *Silica Refractory Brick*

ASTM C530, *Structural Clay Non-Loadbearing Screen Tile*

ASTM C652, *Hollow Brick*

ASTM C902, *Pedestrian and Light Traffic Paving Brick*

ASTM C1261, *Firebox Brick for Residential Fireplaces*

ASTM C1272, *Heavy Vehicular Paving Brick*

UBC 21-1, *Building Brick, Facing Brick and Hollow Brick (Made From Clay or Shale)*

UBC 21-9, *Unburned Clay Masonry Units and Standard Methods of Sampling and Testing Unburned Clay Masonry Units*

Cementitious Masonry Units

ASTM C55, *Concrete Building Brick*

ASTM C73, *Calcium Silicate Face Brick (Sand-Lime Brick)*

ASTM C90, *Loadbearing Concrete Masonry Units*

ASTM C129, *Non-Loadbearing Concrete Masonry Units*

ASTM C139, *Concrete Masonry Units for Construction of Catch Basins And Manholes*

ASTM C744, *Prefaced Concrete and Calcium Silicate Masonry Units*

ASTM C936, *Solid Concrete Interlocking Paving Units*

ASTM C1319, *Concrete Grid Paving Units*

UBC 21-2, *Calcium Silicate Face Brick (Sand-Lime Brick)*

UBC 21-3, *Concrete Building Brick*

UBC 21-4, *Hollow and Solid Loadbearing Concrete Masonry Units*

UBC 21-5, *Non-Loadbearing Concrete Masonry Units*

Natural Stone

ASTM C119, *Terminology Relating to Building Stone*
ASTM C503, *Marble Building Stone*
ASTM C568, *Limestone Building Stone*
ASTM C615, *Granite Building Stone*
ASTM C616, *Sandstone Building Stone*
ASTM C629, *Slate Building Stone*

Mortar and Grout

ASTM C5, *Quicklime for Structural Purposes*
ASTM C33, *Aggregates for Concrete*
ASTM C91, *Masonry Cement*
ASTM C144, *Aggregate for Masonry Mortar*
ASTM C150, *Portland Cement*
ASTM C199, *Pier Test for Refractory Mortar*
ASTM C207, *Hydrated Lime for Masonry Purposes*
ASTM C270, *Mortar for Unit Masonry*
ASTM C330, *Lightweight Aggregates for Structural Concrete*
ASTM C331, *Lightweight Aggregates for Concrete Masonry Units*
ASTM C404, *Aggregates for Masonry Grout*
ASTM C476, *Grout for Reinforced and Nonreinforced Masonry*
ASTM C658, *Chemical Resistant Resin Grouts*
ASTM C887, *Packaged, Dry, Combined Materials for Surface Bonding Mortar*
ASTM C1142, *Extended Life Mortar for Unit Masonry*
ASTM C1329, *Mortar Cement*
UBC 21-11, *Cement, Masonry*
UBC 21-12, *Quicklime for Structural Purposes*
UBC 21-13, *Hydrated Lime for Masonry Purposes*
UBC 21-14, *Mortar Cement*
UBC 21-15, *Mortar for Unit Masonry and Reinforced Masonry Other Than Gypsum*
UBC 21-19, *Grout for Masonry*

Reinforcement and Accessories

ASTM A82, *Cold Drawn Steel Wire for Concrete Reinforcement*
ASTM A153, *Zinc Coating (Hot-Dip) on Iron or Steel Hardware*
ASTM A165, *Electro-Deposited Coatings of Cadmium on Steel*
ASTM A167, *Stainless and Heat Resisting Chromium-Nickel Steel Plate, Sheet and Strip*
ASTM A185, *Welded Steel Wire Fabric for Concrete Reinforcement*
ASTM A496, *Deformed Steel Wire for Concrete Reinforcement*

ASTM A615, *Deformed and Plain Billet-Steel Bars for Concrete Reinforcement*

ASTM A616, *Rail-Steel Deformed and Plain Bars for Concrete Reinforcement*

ASTM A617, *Axle-Steel Deformed and Plain Bars for Concrete Reinforcement*

ASTM A641, *Zinc Coated (Galvanized) Carbon Steel Wire*

ASTM A951, *Joint Reinforcement*

ASTM B227, *Hard-Drawn Copper-Covered Steel Wire, Grade 30HS*

ASTM C1242, *Guide For Design, Selection and Installation of Exterior Dimension Stone Anchors and Anchoring Systems*

UBC 21-10, *Joint Reinforcement for Masonry*

Sampling and Testing

ASTM C67, *Sampling and Testing Brick and Structural Clay Tile*

ASTM C97, *Absorption and Bulk Specific Gravity of Natural Building Stone*

ASTM C109, *Compressive Strength of Hydraulic Cement Mortars*

ASTM C140, *Sampling and Testing Concrete Masonry Units*

ASTM C170, *Compressive Strength of Natural Building Stone*

ASTM C241, *Abrasion Resistance of Stone*

ASTM C267, *Chemical Resistance of Mortars, Grouts and Monolithic Surfacings*

ASTM C426, *Drying Shrinkage of Concrete Block*

ASTM C780, *Preconstruction and Construction Evaluation of Mortars for Plain and Reinforced Unit Masonry*

ASTM C880, *Flexural Strength of Natural Building Stone*

ASTM C952, *Bond Strength of Mortar to Masonry Units*

ASTM C1006, *Splitting Tensile Strength of Masonry Units*

ASTM C1019, *Sampling and Testing Grout*

ASTM C1072, *Method of Measurement of Masonry Flexural Bond Strength*

ASTM C1093, *Accreditation of Testing Agencies for Unit Masonry*

ASTM C1148, *Measuring the Drying Shrinkage of Masonry Mortar*

ASTM C1194, *Compressive Strength of Architectural Cast Stone*

ASTM C1195, *Absorption of Architectural Cast Stone*

ASTM C1196, *In Situ Compressive Stress Within Solid Unit Masonry Estimated Using Flatjack Method*

ASTM C1197, *In Situ Measurement of Masonry Deformability Using the Flatjack Method*

ASTM C1262, *Evaluating the Freeze-Thaw Durability of Manufactured Concrete Masonry and Related Concrete Units*

ASTM C1314, *Method of Constructing and Testing Masonry Prisms Used To Determine Compliance with Specified Compressive Strength of Masonry*

ASTM C1324, *Examination and Analysis of Hardened Masonry Mortar*

ASTM D75, *Sampling Aggregates*

ASTM E72, *Conducting Strength Tests of Panels for Building Construction*

ASTM E447, *Compressive Strength of Masonry Prisms*

ASTM E488, *Strength of Anchors in Concrete and Masonry Elements*

ASTM E514, *Water Permeance of Masonry*

ASTM E518, *Flexural Bond Strength of Masonry*

ASTM E519, *Diagonal Tension in Masonry Assemblages*

ASTM E754, *Pullout Resistance of Ties and Anchors Embedded in Masonry Mortar Joints*

UBC 21-6, *In-Place Masonry Shear Tests*

UBC 21-7, *Anchors in Unreinforced Masonry Walls*

UBC 21-16, *Field Test Specimens for Mortar*

UBC 21-17, *Compressive Strength of Masonry Prisms*

UBC 21-18, *Sampling and Testing Grout*

UBC 21-20, *Flexural Bond Strength of Mortar Cement*

Assemblages

ASTM C901, *Prefabricated Masonry Panels*

ASTM C946, *Construction of Dry Stacked, Surface Bonded Walls*

ASTM E835, *Guide for Dimensional Coordination of Structural Clay Units, Concrete, Masonry Units, and Clay Flue Linings*

ASTM C1283, *Practice for Installing Clay Flue Lining*

ASTM E1602, *Guide for the Construction of Solid Fuel-Burning Masonry Heaters*

UBC 21-8, *Pointing of Unreinforced Masonry Walls*

The NCCER makes every effort to keep these manuals up-to-date and free of technical errors. We appreciate your help in this process. If you have an idea for improving this manual, or if you find an error, a typographical mistake, or an inaccuracy in the NCCER's Craft Training Manuals, please write us, using this form or a photocopy. Be sure to include the exact module number, page number, a description of the problem, and the correction, if possible. Your input will be brought to the attention of the Technical Review Committee. Thank you for your assistance.

Instructors – If you found that additional materials were necessary in order to teach this module effectively, please let us know so that we may include them in the Equipment/Materials list in the Instructor's Guide.

Write: Curriculum and Revision Department
National Center for Construction Education and Research
P.O. Box 141104
Gainesville, FL 32614-1104

Fax: 352-334-0932

Craft _____   Module Name _____

Module Number _____   Page Number(s) _____

Description of Problem

_____

_____

_____

_____

(Optional) Correction of Problem

_____

_____

_____

(Optional) Your Name and Address

_____

_____

_____

Mortar

Module 28105

NATIONAL
CENTER FOR
CONSTRUCTION
EDUCATION AND
RESEARCH

OBJECTIVES

Upon completion of this module, the trainee will be able to:

1. Name and describe the primary ingredients in mortar and their properties.
2. Identify the various types of mortar used in masonry work.
3. Describe the common admixtures and their uses.
4. Identify the common problems found in mortar application and their solutions.
5. Properly set up the mortar mixing area.
6. Properly mix mortar by hand.
7. Properly mix mortar with a mechanical mixer.

Prerequisites

Successful completion of the following Task Modules is recommended before beginning study of this Task Module: Core Curricula; Masonry Level 1, Modules 28101 through 28104.

Required Trainee Materials

1. Trainee Task Module
2. Pencil and paper
3. Appropriate Personal Protective Equipment

COURSE MAP

This course map shows all of the task modules in the first level of the Masonry curricula. The suggested training order begins at the bottom and proceeds up. Skill levels increase as a trainee advances on the course map. The training order may be adjusted by the local Training Program Sponsor.

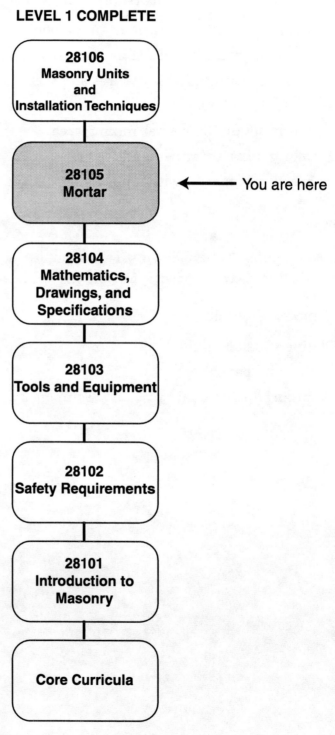

LEVEL 1 COMPLETE

28106
Masonry Units
and
Installation Techniques

28105
Mortar

← You are here

28104
Mathematics,
Drawings, and
Specifications

28103
Tools and Equipment

28102
Safety Requirements

28101
Introduction to
Masonry

Core Curricula

Cmap105.EPS

TABLE OF CONTENTS

Trade Terms Introduced In This Module

Admixture: A chemical or mineral other than water, cement, or aggregate added to mortar immediately before or during mixing to change its setting time or curing time, to reduce water, or to change the overall properties of the mortar.

Air-entraining: A type of admixture added to mortar to increase microscopic air bubbles in mixed mortar. The air bubbles increase resistance to freeze-thaw damage.

Hydration: A chemical reaction between cement and water that hardens the mortar. Hydration requires the presence of water and an air temperature between 40° and 80° F.

Masonry cement: Cement that has been modified by adding lime and other materials.

Plasticity: The ability of mortar to flow like a liquid and not form cracks or break apart.

Slaked lime: Lime reduced by mixing with water to a safe and usable form that can be used in the production of mortar.

Water retention: The ability of mortar to keep sufficient water in the mix to enhance plasticity and workability.

Workability: The property of mortar to remain soft and plastic long enough to allow the mason to place and align masonry units and strike off the mortar joints before the mortar hardens completely.

1.0.0   INTRODUCTION

Mortar is one of the basic building materials used by the mason. Like other materials used in masonry construction, mortar can be mixed to meet a variety of situations. This module introduces the basic materials used to make mortar, the types of mortar used, how to mix mortar, and its characteristics.

Masonry structures are a combination of mortar and masonry units. Since mortar bonds the masonry units together, it is critical to the strength of the overall structure. Mortar serves to:

*   Bond the units together, sealing the spaces between them
*   Make up the differences in the size of the units (bricks and concrete blocks)
*   Bond metal ties, grids, and anchors
*   Make the structure airtight and watertight
*   Create a neat, good looking, uniform appearance

Masons should be able to mix mortar and describe how each of the ingredients contributes to the mix. The mason should also be able to recognize the difference between good and poor mortar.

2.0.0 MORTAR MATERIALS

In its simplest form, mortar is made up of cement, lime, sand, and water. Other materials are added to achieve various physical or chemical characteristics needed for a particular application. The basic building blocks of mortar will be presented in this section. The next paragraphs discuss the cement, lime, sand, and water that go into mortar.

2.1.0 PORTLAND CEMENT

Mortar is mostly portland cement. Portland cement gives mortar its characteristic color and many of its properties. Both mortar and cement have a limited workable life span, both need similar safety precautions, and both achieve their strength by the same chemical process of **hydration**.

Mortar differs from concrete in several ways. Mortar binds masonry units into a single component, while concrete is usually a structural element itself. Mortar is often mixed by hand, while concrete is usually mixed by machine on the job or is brought to the job site already mixed. Mortar is placed and manipulated by hand considerably more often than concrete. Therefore, it must maintain its **plasticity** and **workability** longer than concrete. For these reasons, mortar is mixed, treated, and used in a vastly different manner than concrete.

ASTM specifications list five different types of portland cement which may be used for mortar. Generally, only Types I, II, and III are used for mortar while the other two are usually used only for concrete.

Type I   This is a general-purpose cement and is the one masons use most often. It may be used in pavements, sidewalks, reinforced concrete bridge culverts, and masonry mortar.

Type II   (Modified) This cement hydrates at a lower temperature than Type I and generates less heat. It does, however, have better resistance to sulfate than Type I. It is usually specified for use in such places as large piers, heavy abutments, and heavy retaining walls.

Type III   (High early strength) Although this cement requires as long to set as Type I, it achieves its full strength much sooner. Generally, when high strength is required in one to three days, this cement is recommended. In cold weather, when protection from freezing weather is important, Type III is specified often.

Type IV   (Low heat) This is a special cement for use where the amount and rate of heat generated must be kept to a minimum. It is critical to hold the temperature down to ensure that the concrete cures properly. Since the concrete cures slowly, strength also develops at a slower rate. Low heat portland cement is used in areas where there are large masses of concrete, such as dams or large bridges, and where Type I or Type II cement would generate too much heat.

Type V (Sulfate resistant) This is intended for use only in construction exposed to severe sulfate actions. It also gains strength at a slower rate than normal portland cement.

2.2.0 HYDRATED LIME

The second ingredient in mortar is a special type of lime called hydrated lime. Hydrated lime is quicklime that has been **slaked**, which means that the lime has been mixed with water to form a putty. This process takes several weeks and is potentially dangerous, since considerable heat is released by the lime/water reaction. Manufacturers of hydrated lime slake the lime as part of producing lime for mortar.

Two types of hydrated lime are made: Type S and Type N. Only Type S lime is recommended for mortar.

Hydrated lime adds several desirable characteristics to mortar.

- *Workability* – This is the degree of smoothness and malleability of the mortar. Lime helps mortar to be very workable, enabling the mason to butter joints without great effort. This results in better workmanship and more economical construction. Mortar with the correct lime content spreads easily with the trowel; cement which is not mixed with lime is more difficult to spread.

- *Water retention*: Lime increases the water-holding capacity of the mortar, decreases the loss of water from the mortar (bleeding), and reduces the separation of sand. Mortar with a low lime content soon looses its moisture and sets prematurely when spread on the wall. If the mason does not have enough time to lay the masonry units before the mortar drys out, a poor bond will result, causing a weak structure.

- *Bond strength* – Bond is probably the most important property of mortar. Although the addition of lime to mortar decreases the strength of the mortar, it helps the mortar fill gaps and adhere to the structure, resulting in a higher bond strength.

- *Flexibility* – For mortar to withstand strong winds, lateral pressures, and hard jolts, it must have enough flexibility to prevent cracking. Tall masonry structures, such as high chimneys, may sway as much as 12 inches. The addition of lime increases the plasticity and flexibility of the mortar.

- *Resistance to weather* – Mortar should be able to resist strong winds, freezing temperatures, and alternating wet and dry weather. Wetting and drying cycles actually help lime-based mortar to increase in overall strength as it ages. In fact, when small cracks or separations appear in the mortar joints, the action of lime combined with rainwater often reseals these cracks.

- *Economy* – In addition to being inexpensive, lime helps make a smoother, more uniform mortar. Lime-based mortars have a slower set time which offers economy by reducing the amount of retempering.

- *Minimal shrinkage* – Very hard, high strength mortar will sometimes crack because it shrinks after hardening. These cracks result in a loss of strength. Lime undergoes the second least loss in volume of all the ingredients in mortar.

- *Autogenous healing* – The ability of lime to reseal itself when hairline cracks form in the mortar is called autogenous healing. Lime reacts with water and elements in the air to reseal these minute cracks. This process is the result of a chemical process called recarbonation.

2.3.0 MASONRY CEMENT

Most mortars are made from portland cement which must be mixed with other materials. In recent years, **masonry cements** have been produced that require only the addition of sand and water to make mortar.

Masonry cement mortar mixes provide an all-purpose mortar containing all the ingredients except sand and water in one bag. Most masonry cements are a combination of portland cement, lime, gypsum, and **air-entraining admixtures**. A bag of masonry cement usually weighs 70 pounds. The quality of the mix and its appearance are normally very consistent because the materials in the masonry cement are ground together and mixed before packaging.

Masonry cement is less subject to variations from batch to batch than mortar produced by mixing all the ingredients on the job. However, sand, which is the most varying material, must be added to both types on the job. When masonry cement is used, there are fewer materials to handle and mixing is considerably faster. Masonry cement is made in grades M, S, N, and O. Masonry cements have the merit of economy and convenience, but care should be taken in using these products. Some tests show they will not meet the required specifications for all types of mortar.

2.4.0 SAND

Sand is a primary ingredient in mortar. There are two types of sand available for making mortar: manufactured and natural. Manufactured sand comes from crushing stone, gravel, or air-cooled blast furnace slag. It has a sharp, angular grain and produces mortar with different properties than natural sand. Natural sand found on lake bottoms, river beds, or river banks is rounder and smoother than manufactured sand. Seashore sand has a very high salt content and should never be used in mortar.

Sand acts as a filler in mortar and adds to the strength of the mix. It also decreases shrinkage during the hardening phase and, therefore, decreases cracking. The water, cement, and lime in the mix form a paste that coats the sand particles and lubricates them to form a workable mix. The amount of air contained in the mix is determined by the range of particle sizes and grades of sand used. *Figure 1* shows the recommended gradation limits for sand. The percentages show how much of the sand must pass through a set of screens with progressively finer mesh sizes. Note that 35 to 70 percent of the sand must pass through the #30 screen, which has finer mesh than the #4 screen.

	Percent Passing	
Sieve Size	Natural Sand	Manufactured Sand
No. 4	100	100
No. 8	95–100	95–100
No. 16	60–100	60–100
No. 30	35–70	35–70
No. 50	15–35	20–40
No. 100	2–15	10–25
No. 200	–	0–10

105F01.EPS

Figure 1. Recommended Sand Gradation Limits

It is important to use good quality sand in the production of mortar. Sand is one of the least expensive building materials on the job. Cutting costs by using poor quality sand will lead to a variety of problems. If silt is present in mortar, the mortar will stick to the masons' trowels, causing a drop in their production rate. Silt reduces the strength of the mortar by preventing the cementitious material from bonding to the sand. Silt will cause small mud pits and holes to form on the surface of the joints. The mortar will not be of a uniform color and the amount of impurities will vary from batch to batch. Poor quality sand has no place on the job as it lowers both the quality of the final product and the mason's productivity.

If impurities are suspected in sand, a simple siltation test can determine if the sand is satisfactory for use in masonry mortar. The test is a simple process that can be done at the job site in these steps:

Step 1 Fill a quart glass jar half full of the sand and add water to a level three inches above the top of the sand. Place the cap on the jar and shake it vigorously.

Step 2 Set the jar aside until the next morning. The sand will settle to the bottom of the jar and any foreign material will come to rest on top of the sand.

Step 3 If the accumulated silt and organic material measures more than ⅛ of an inch, the sand should not be used. *Figure 2* shows an unacceptable result of a sand siltation test.

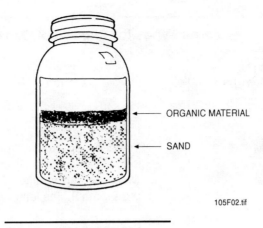

ORGANIC MATERIAL

SAND

105F02.tif

Figure 2. Siltation Test Jar

2.5.0 WATER

Water used in mortar must be free of all chemicals and foreign matter that would weaken or discolor the mix. These chemicals include alkalis, salts, acids, and any organic materials. Generally, water that is safe to drink is acceptable for use with mortar. Water purification chemicals found in city water supplies usually have no adverse effects on mortar. Water should be used at a cool ambient temperature.

3.0.0 MORTAR TYPES

Several different types of mortar have been developed to meet the needs of different job requirements. Selecting the right mortar type will depend on the type of masonry work and its location in the structure.

Five types of mortar are recommended for various classes of masonry construction. These five classes are designated by the letters M, S, N, O, and K and are the same as that used for masonry cement. These letters can be remembered by taking every other letter from the words "mason work." *Figure 3* displays the specifications for mixing each type of mortar. *Figure 4* shows the type of construction suitable for each mortar.

Mortar Type	Parts by Volume of Portland Cement	Parts by Volume of Hydrated Lime	Sand, Measured in a Damp, Loose Condition
M	1	$1/4$	
S	1	$1/2$	Not less than $2\frac{1}{4}$ and not more than 3 times the sum of the volumes of cement and lime used.
N	1	1	
O	1	2	
K	$1/4$	1	

105F03.EPS

Figure 3. Proportion Specifications For Mortar

Mortar Types for Classes of Construction	
ASTM Mortar Type Designation	**Construction Stability**
M	Masonry subjected to high compressive loads, severe frost action, or high lateral loads from earth pressures, hurricane winds, or earthquakes. Structures below grade, manholes, and catch basins.
S	Structures requiring high flexural bond strength, but subject only to normal compressive loads.
N	General use in above grade masonry. Residential basement construction, interior walls, and partitions. Concrete masonry veneers applied to frame construction.
O	Nonloadbearing walls and partitions. Solid loadbearing masonry of allowable compressive strength not exceeding 100 psi.
K	Interior nonloadbearing partitions where low compressive and bond strengths are permitted by building codes.

105F04.EPS

Figure 4. Mortar Types For Classes Of Construction

- Type M mortar has the highest compressive strength and is somewhat more durable than the other mortars. In laboratory tests, the ASTM has found that a 2-inch cube of Type M mortar will stand up to at least 2500 psi of compression after it has cured for 28 days.

- Type S mortar has a medium compressive strength, good workability, and excellent durability. This type is often interchanged with Type M because it offers a relatively high compression strength but better workability, **water retention**, and flexibility. The ASTM specifications state that Type S mortar must stand up to at least 1800 psi of compression.

- Type N mortar is a medium-strength mortar with a psi rating of at least 750. It has excellent workability, although it does not have the strength of Types M or S. Type N is the most frequently used mortar because it is highly resistant to weather.

- Type O mortar is a low-strength mortar with very good workability. Because of its low durability, this type is not recommended for use where bad weather conditions exist. Type O mortar should have a compression strength of at least 350 psi.

- Type K mortar is so low in compressive strength and in bond strength that it is seldom used. The ASTM requires that the testing strength be only 75 psi after 28 days of curing. Type K mortar made without cement is sometimes used for restoration of historic buildings.

4.0.0 MORTAR PROPERTIES

Properties of mortar can be divided into two categories related to its physical state. The first state begins immediately after it is mixed and is in use on the job. This period is when the mortar is in its plastic or workable state. The second state begins after the mortar has cured into its final rigid condition; this is called the hardened state. As a mason you will be primarily concerned with the plastic state of mortar since that is when you will be working with it. Since the mortar quickly moves to the hardened state and remains there for the life of the structure, you must also be familiar with the properties of mortar in its hardened state.

4.1.0 PLASTIC PROPERTIES

Plastic mortar is freshly mixed and ready for use on the job. The plastic properties of mortar include its workability, water retention, water content, and rate of hardening.

4.1.1 Workability

This property of mortar is difficult to define as there are several factors that influence workability. The properties having the most influence on workability are consistency, setting time, adhesion, and cohesion.

- Consistency or uniformity is the indicator of how similar mortar is from one batch to the next and from one place to another in the same batch. Consistency is controlled and improved by closely following the mix design and the mixing procedure. This includes the order in which the ingredients are added as well as the mixing time.

- Setting time establishes how much time the mason will have to use the mortar before it begins to harden to the point of not being workable. Setting time is determined primarily by the type of cement used in the mix. Setting times can be extended by using the maximum water allowable in the mix, by wetting the mortar board prior to placing the mortar on it, and by tempering the mortar during its use.

- Adhesion is the indicator of how well the mortar sticks to the masonry units. Greater adhesion means greater bonding strength.

- Cohesion is the indicator of how well mortar sticks to itself. Mortar with good cohesion will extrude between masonry units without smearing or dropping away.

4.1.2 Water Retention

Water retention is related to workability. Mortar having good water retention remains soft and plastic long enough for the mason to align, level, and plumb the blocks without danger of breaking the contact or bond between the mortar and the units.

Sufficient water retention is essential if the mortar is to develop the necessary adhesion between itself and the masonry units to make strong and watertight joints. Mortar must also maintain sufficient water content to allow the cement to hydrate. This is the chemical process which transforms the plastic mortar into its hardened state.

Water retention is improved by the addition of more lime, finer sand, entrained air, and more water. These also increase the workability of the mortar. However, care must be taken that the proportions of the ingredients continue to stay within the limits specified in the mix formula.

4.1.3 Water Content

Water content is possibly the most misunderstood aspect of masonry mortar. This is probably due to the similarity between mortar and concrete materials. Many designers have mistakenly based mortar specifications on the assumption that mortar requirements are similar to concrete requirements, especially with regard to the water-cement ratio. *Figure 5* shows a comparison of the typical stiffness for concrete, mortar, and grout. To test the materials, the cone is filled and inverted. The stiffer materials will keep the shape of the cone better than the more plastic materials.

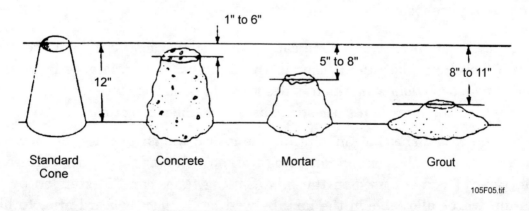

105F05.tif

Figure 5. Stiffness Test Comparison

Many specifications incorrectly require mortar to be mixed with the minimum amount of water consistent with workability. Often retempering of mortar is also prohibited. These provisions result in mortars that have higher compressive strengths but lower bond strengths. Mixing mortar with the maximum amount of water consistent with workability will provide maximum bond strength within the capacity of the mortar. Retempering is permitted, but only to replace water lost by evaporation. This can usually be controlled by requiring that all mortar be used within 2½ hours after mixing, depending on conditions.

4.1.4 Consistent Rate Of Hardening

Mortar hardens when cement reacts chemically with water in a process called hydration. The rate of hardening of mortar is the speed at which it develops resistance to an applied load. Too rapid hardening may interfere with the use of the mortar by the mason. Very slow hardening may impede the progress of the work since the mortar will flow from the completed masonry. During winter construction, slow hardening may also subject mortar to early damage from frost action. A well defined, consistent rate of hardening assists the mason in laying the masonry units and in tooling the joints at the same degree of hardness. Uniform joint color of masonry reflects proper hardening and consistent tooling times.

4.2.0 HARDENED PROPERTIES

Properties of hardened mortar are the result of the characteristics and quality of the materials used in the mix and of the workmanship used by the mason.

4.2.1 Durability

Durability of mortar is measured primarily by its ability to resist the effects of repeated cycles of freezing and thawing under natural weather conditions. High compressive strength mortars usually have greater durability due to their higher density. Air-entrained mortars also withstand freeze-thaw cycles better than mortars without air-entraining admixtures. Mortars made with masonry cement have higher air content than mortars made with portland cement and lime and, therefore, resist freeze-thaw cycles better.

Sulfate resistance is another factor affecting durability of mortars. The use of sulfate-resistant material should be considered when mortars will be in contact with soil, groundwater, or industrial processes. Type V portland cement is sulfate-resistant and can be used in these situations, but Type V cement is slow to gain strength. If sulfate-resistant materials are not used in the mortar mix, then a protective coating can be applied to the finished work.

4.2.2 Compressive Strength

Concrete masonry structures gain their strength from the compressive strength of the masonry units, the proportions of materials used in the mortar mix, the design of the structure, the workmanship of the mason, and the amount of curing.

The compressive strength of mortar increases as the cement content is increased, and decreases as more lime, air entrainment, and water are added.

4.2.3 Mechanical Bond

Bond strength is the ability of the mortar to hold the masonry units together. Mechanical bond strength depends on the adhesion of the mortar to the surface of the masonry unit and on the amount of contact between the mortar and the masonry unit.

Spreading the mortar completely over the unit face or shell is essential, and the mortar must have enough flow and workability to wet the contact surfaces. Masonry units have surface irregularities that increase mechanical bond by increasing the area of the surface to be bonded. Both clay and concrete products have enough surface absorption to draw the wet mortar into their irregularities. *Figure 6* shows an enlargement of this process.

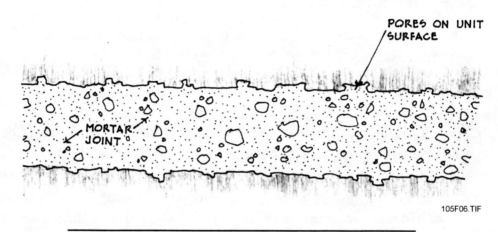

Figure 6. Mortar Bonding To A Porous Masonry Surface

Mechanical bond strength in cured mortar depends on the mechanical interlocking of cement hydration crystals formed in the masonry unit pores and on its surface. Voids, air holes in the mortar joint, weaken the mechanical bond. They also allow water into the joint. Smooth masonry units such as molded brick or smooth stone do not have these micropores, so they provide less mechanical bond with the mortar. Loose sand particles, dirt, and other contamination also weaken the mechanical bond.

Skill is critical in bonding. Full mortar joints must have complete contact on the appropriate surfaces. Once it is in place, a unit will break or weaken its bond if it is moved. Wet mortars will allow more time for placing units. Laboratory tests show that tapping the unit to level increases bond strength 50 to 100 percent over hand pressure alone.

The strength of the bond depends on the strength of the mortar itself. The main factors affecting mortar strength are:

• Ingredients, such as types and amounts of cement, retained water, and air content

• Surface texture and moisture content of the masonry units used

• Curing conditions, such as temperature, wind, relative humidity, and amount of retained water

Mortar strength is directly related to cement content. Both bond strength and mortar stiffness increase as cement content increases. Bond strength also increases as water content increases. The optimum bond strength is obtained by using a mortar with the highest water content compatible with workability.

4.2.4 Volume Change

Volume changes in mortar are of two types: shrinkage, which causes cracks to appear in joints; and expansion, which is due to poor quality ingredients.

Shrinkage across a mortar joint made with properly proportioned mortar is usually very small and not of concern. Shrinkage cracks can become a problem when there is a large amount of water loss from the mortar. The rate of water loss from mortar is determined by several factors. Primary among these factors are the amount of water in the mortar and the rate at which it will be absorbed into the masonry units. Air-entrained mortars need less water than mortars which are not air-entrained. It should be remembered, however, that workability and bond strength are directly related to the flow of the mortar and should be given priority when determining water content in the mixing operation.

Expansion in mortars is generally caused by using poor quality material in the mixing process. Attention to the storage and handling of the material will eliminate much of this problem. There are tests that can be run to check for poor quality. The siltation test described earlier will assure that the sand is clean. The autoclave expansion test (ASTM C91) measures the quality of the cementitious material.

4.2.5 Appearance

Uniformity of color and shade of the mortar joints greatly affects the overall appearance of a masonry structure. Atmospheric conditions, admixtures, and moisture content of the masonry units are some of the factors affecting the color and shade of mortar joints. Other factors are uniformity of proportions in the mortar mix, water content, and time of tooling the mortar joints.

Careful measurement of mortar materials and thorough mixing are important to maintain uniformity of color and shade.

5.0.0 MORTAR ADMIXTURES

Many materials or additives can be put into mortar to change its appearance or properties. Admixtures are additives put into cement before it is mixed with lime and sand to form mortar. *Figure 7* lists admixtures commonly used with cement and their effects on mortar.

Admixture	Workability	Strength	Weather Resistance
Air-Entraining Agents	Increase	Decrease slightly	Increase
Bonding Agents	–	Increase	–
Plasticizers	Increase	–	–
Set Accelerators	Decrease	Increase	–
Set Retarders	Increase	–	–
Water Reducers	Decrease	Increase	Increase
Water Repellents	–	–	Increase
Pozzolanic Agents	Increase	Increase	–

105F07.EPS

Figure 7. Admixtures And Their Effects

- Air-entraining admixtures reduce water content and increase workability and freeze-thaw resistance. Bond strength may be reduced slightly. If air entraining is needed, it is recommended to use cements modified at the factory to include an air-entraining agent.

- Bonding admixtures increase the adhesion of masonry mortar to masonry units. Some organic modifiers provide an air-cure adhesive that increases bond strength of dry masonry.

- Plasticizers increase workability. Inorganic plasticizers, such as clay, clay-shale, and finely ground limestone, promote workability and water release for cement hydration. Organic plasticizers also promote workability; however, mortar may stick more to the mason's tools.

- Set accelerators stimulate cement hydration and increase compressive strength of laboratory test mortars at normal temperatures. Nonchloride accelerators are preferred because chloride salts will deteriorate the cement in masonry.

- Set retarders delay cement hydration, providing more time for the mortar to remain plastic without retempering. This is useful primarily during hot weather. Set retarders are also used in marketing ready-mixed, set-controlled masonry mortars. The effect of the modifier persists for up to 72 hours but dissipates when retarded mortar contacts absorptive masonry units.

- Water reducers lower the amount of water and potentially increase mortar strength. However, masonry mortar rapidly loses water in contact with an absorptive masonry unit. Water reducers may reduce the water below the level needed for cement hydration. Care must be taken to insure sufficient water for the hydration process to take place.

- Water repellents modify masonry mortar during the early period after construction. If persistent in their performance, rain penetration of the masonry may be reduced.

- Pozzolanic modifiers, because of their fineness and ability to combine with lime, will cause mortar to increase density and strength.

- Antifreeze admixtures are not recommended for use in mortar as the amounts needed to prevent freezing also significantly lower the compressive and bonding strength of mortar.

6.0.0 COLORANTS

Architectural effects can be obtained with mortar colored to contrast or harmonize with the masonry units. White mortar is obtained by using white portland cement with lime and white sand. If pastel colors such as cream, ivory, pink, and rose are desired, then the use of white mortar is essential. Darker colors can be achieved by using colorants in normal gray mortar. Colors may not be as bright with a gray mortar base, but the result is much less costly.

Pigments used to achieve the coloring must be adequately dispersed throughout the mix in order to be effective. To determine if mixing is sufficient, take a small sample of the mortar and flatten it out under a trowel. If there are streaks present, additional mixing is needed. Prepackaged colored masonry cement is the easiest and best way to achieve color results. Colored masonry cements are available in many areas as types M, S, and N.

As a rule, color pigments should be composed of mineral oxide and contain no dispersants that will slow or stop the hydration process. Mineral oxides should not exceed 5 percent by weight of cement for masonry cement or 10 percent by weight of cement for cement-lime mortars. Iron, manganese, chromium, and cobalt oxides have also been used successfully. Zinc and lead oxides should be avoided because they may react with cement. Carbon black may be used as a coloring agent to obtain dark gray or almost black mortar, but lampblack should not be used. Carbon black should be limited to 1 percent or 2 percent by weight of cement for masonry cement or cement-lime mortars, respectively. The color of carbon black mortar rapidly fades with exposure to weathering. Carbon black mortar has durability problems as well.

It is recommended that only those pigments found acceptable by testing and experience be used. The following is a guide to the selection of coloring material:

- Red, yellow, brown, black, or gray colors should come from iron oxide pigments.
- Green coloring should be made with chromium oxide pigments.
- Blue coloring should be made with cobalt oxide pigments.

Only the smallest amount needed to achieve the desired color should be used. If pigment is in excess of 10 percent of the portland cement by weight, the mortar may have problems in strength and durability.

7.0.0 SETTING UP, STORING, AND MEASURING

Masons should be able to mix their own mortar, or "mud" as it is known in the trade. They need to know how to modify the mix for different requirements of the work. On large jobs, mortar may not be mixed on the site but brought in from an off-site mixing plant. On small jobs, however, it may be practical for masons to mix their own mortar either by hand or with a power mixer. The following sections present the steps needed to prepare for mixing mortar.

7.1.0 SETTING UP THE MORTAR AREA

Efficient placing of the materials and the mixer can save time and work. Because they need to be close together, the mixer site should be set at the same time as materials stockpile sites are set. Avoid a site that is downhill from the work area because the mortar will be moved in wheelbarrows.

The mixer should be located as close as possible to the main section of the work with mix ingredients arranged around it. The stockpiles should be on the side of the mixer away from the work so that the mortar-filled wheelbarrows do not have to pass around the stockpiles. If the ground is soft or damp, set up a plywood runway for the filled wheelbarrows.

A hose can be used to bring the water to a barrel near the mixer or to fill the water measure directly. Make sure the hose does not cross the wheelbarrow path. Keep the area clean by measuring and loading materials carefully. Clear away spills with a shovel or brush. A clean mixing site is safer as well as easier to use.

Setting up for a good-sized job, the mortar mixing area needs to be:

- away from materials and equipment movement paths
- close to stockpiles so materials will not have to be moved a long distance
- close to the work site so mortar will not have to be moved a long distance, but not so close as to interfere with the work
- not downhill from the work area, so mortar-filled wheelbarrows will not have to be pushed uphill

If the job is a small job requiring only small amounts of mortar, the mortar can be mixed in a wheelbarrow and rolled directly to the work site.

7.2.0 STORING MATERIALS

The materials used in making mortar should be stored on the job site in an area convenient to the mixing site, but not where it will interfere with the work. Cement and lime are generally sold and delivered by the bag, while sand is ordered by weight but delivered in loose form.

For this reason, they have somewhat different storage requirements. The cement and lime should be kept in a storage shed. If kept outside, they must be covered with plastic sheets or canvas tarpaulins. In either case, the bags should be kept off the ground by using wooden pallets to keep them from ground moisture. *Figure 8* shows blocks covered to protect them from moisture. Any other materials stored outside should be covered in a similar manner.

105F08.tif

Figure 8. Storing Blocks

Sand can be dumped on the ground and does not need to be protected from moisture in warm weather. In winter, the stockpile should be covered to prevent wet sand from freezing. Another method to keep sand from freezing is to run pipes through the stockpile, then pump heated air or water through the pipes. If the sand does freeze, it must be thawed before it can be used. Care should also be taken to keep foreign material out of the stockpile, especially where the sand has been dumped on muddy or vegetation-covered ground. If the only location for sand at the job site is muddy or vegetation-covered ground, the sand can be dumped on plastic sheets or tarpaulins.

7.3.0   MEASURING MORTAR MATERIALS

Quality control of a mortar mix and consistency between batches starts with the proper measuring of mortar ingredients. Proper measuring will ensure that the properties of the mortar—workability, color, strength, and so forth—will be the same from mix to mix. This ability to produce the same results in each batch will increase the mason's productivity and lead to a quality, good-looking job.

Mortar mix specifications are expressed in terms of a proportion of cement, lime, and sand volumes. For example, for a Type N mortar, the proportions might be expressed as 1:1:6, or 1 part portland cement, 1 part lime, and 6 parts sand. For a Type O cement-lime mortar with a mix ratio of 1:2:9, the proportions would be 1 part portland cement, 2 parts lime, and 9 parts sand.

Portland cement is available in 94-pound bags containing one cubic foot of material. Hydrated lime comes in 50-pound bags containing one cubic foot. Masonry cement comes in 70-pound bags containing one cubic foot. Since sand comes in bulk form, it needs some practical and consistent method of measuring it. Many masons measure sand by the shovel. The number of shovels of sand in a cubic foot will vary, however, depending on the amount of water in the sand. One method that works well is to use a cubic foot box as shown in *Figure 9*. This box measures 1 foot long × 1 foot wide × 1 foot deep and holds exactly 1 cubic foot of material.

105F09.tif

Figure 9. Cubic Foot Box

This cubic foot box can be used to measure all the sand used in the mix. It is more commonly used to calibrate the shovel for loading the sand into the mixer. The mason determines the number of shovels of sand required for a cubic foot by filling the box. The box is no longer necessary for most of the mixing operations. You should recheck your shovel count twice each day to allow for factors such as moisture content of the sand, changing shovels, or having different people doing the shoveling.

Measuring the water should always be done with a bucket of some sort. Never pour water directly into the mixer from a hose as it is not possible to calculate the exact amount added. A plastic 5-gallon bucket works well for this purpose.

The volume of mortar obtained from a mix will only equal the volume of sand in the mix, even though you have also added a quantity of cement, lime, or masonry cement. This is because the cement, lime, and water occupy the voids between the grains of sand. The mix becomes more dense as more cement and lime are added but not more voluminous. For this reason, consider only the amount of sand added when you want to know how much mortar will be produced in your mix.

8.0.0 MIXING BY MACHINE

Machine mixing is the preferred method for preparing a batch of mortar. For larger jobs, a machine mixer can produce 4 to 7 cubic feet of mortar at one time with the same quality and the same properties batch after batch. Machine mixing requires less human energy which frees masons to use their energy where it counts most— in the construction of their project.

When using a power mixer, either electric or gasoline-powered, place the materials near the mixing area. The mixer should be at the center of these materials in such a way that all materials can be easily reached and the path for the wheelbarrow is clear. The mixer should be securely supported and the wheels blocked to avoid tire wear. The following sections describe the steps that should be followed when using a power mixer.

CAUTION: Wear eye protection and other Appropriate Personal Protective Equipment when using a power mixer.

8.1.0 MACHINE MIXING STEPS

Review the operation manual for the machine used and review safety procedures for working with power equipment. Have the mix formula written down, including the number of cubic feet of cement and lime as well as the cubic feet or shovelsful of sand needed for the size of the batch to be mixed. Follow these steps for mixing mortar:

Step 1   With the blades turning, add a small amount of water. Add enough to wet the inside of the mixing drum to prevent the mortar from caking on the mixing paddles or the sides of the drum. *Figure 10* shows this step.

105F10.tif

Figure 10. Add Water To The Mixer

Step 2   Add one third to one half of the sand needed for the batch, as shown in *Figure 11*. Keep the paddles turning to prevent stress on the turning mechanism.

105F13.tif

Figure 11. Add Sand To The Mixer

Step 3   Add all the necessary cement and lime or the masonry cement to the mixer, as shown in *Figure 12*. This is best done by the bag. Place the bag on the safety grate and open the bag by cutting with a small knife or trowel, or by pulling the bag opening, if available. Some mixers have a sharp set of metal teeth on the safety grate specifically designed for this purpose.

105F12.tif

Figure 12. Add The Cementitious Materials

MASONRY — TRAINEE TASK MODULE 28105

Step 4 Add the remaining sand at this time, as shown in *Figure 13*. Then add additional water to bring the mixture to the desired consistency. Allow the mixing process to continue for 3 to 5 minutes in order to completely blend the materials. Do not mix any longer, however, as this will allow excessive air to be trapped in the mix. This extra air will cause the mortar to be spongy and weak.

Figure 13. Add Remaining Sand

Step 5 When the mixing is completed, with the drum still turning, grasp the drum handle and dump the mortar into the wheelbarrow or mortar box. *Figure 14* shows the mason catching the mortar in a wheelbarrow. The turning blades will clear most of the mortar out of the mixer. Next, take the blades out of gear and turn off the mixer. The remaining mortar can now be removed by hand.

Figure 14. Dump The Completed Batch

Step 6 If more mortar is needed immediately, start the process over again by adding enough water to the drum to clean the blades as they are turning. If no more mortar will be needed for some time, leave the mixer turned off and clean the inside and the blades with a water hose and a stiff bristle brush, as shown in *Figure 15*.

105F15.tif

Figure 15. Rinse The Drum With Water

8.2.0 SAFETY TIPS FOR POWER MIXERS

As with any piece of power equipment, a power mortar mixer should be treated with respect and caution. The manufacturer's manual should be carefully read, and the procedures listed should be followed. Here are some basic safety tips:

• Wear safety glasses or goggles to protect your eyes, and a nose mask to prevent you from breathing any cement or lime dust.

• When adding the water, shoveling sand, or pouring the cement or lime into the drum, be extremely careful not to place the shovel or bucket into the mouth of the mixer where the blades are turning. Serious injury could be caused by an uncontrolled shovel handle.

• The same caution concerns your hands and clothing: do not place them into the turning mixer.

CAUTION: Keep your hands out of the mouth of the mixer. If a torn bag falls into the mixer, do not try to remove it while the mixer is turning.

• If the mixer does not have a safety grate over the mixing chamber, consider adding one.

- Do not scrape the last mortar out of the drum while the blades are still turning. Disengage the blades and turn off the mixer.

- Final cleanup of the mixer is done only with the mixer turned off.

9.0.0 MIXING BY HAND

For small jobs, manual mixing of the mortar usually makes the most sense. The preliminary steps are the same as those for mechanical or power mixing. Position the material near the mortar box. Ensure that the mortar box is level and stable, as shown in *Figure 16*. Have the mix design written down. Also, ensure that there is enough room at the two ends of the mortar box for you to stand while mixing the mortar. No particular height for the box is required. It should be at a height that will be convenient for the mason when using the hoe to mix the mortar.

105F16.tif

Figure 16. Level And Adjust The Height

Here are the basic steps for mixing mortar by hand:

Step 1 Place one half of the sand in the box, as shown in *Figure 17*. Spread this sand out evenly across the bottom of the box.

105F17.tif

Figure 17. Add Half The Sand

Step 2   Place the desired amount of cement, lime, or masonry cement over the sand, as shown in *Figure 18*. For small batches of mortar, you may wish to use standard shovelsful of material rather than bags as your measuring device. Use a shovel to spread out the remaining sand across the cement and lime layer.

Figure 18. Add Cement And Lime

Step 3   Blend the dry ingredients together with the shovel or hoe. When they are thoroughly mixed, push them to one end of the box, as shown in *Figure 19*.

Figure 19. Push The Mixed Ingredients To One End

Step 4   Add half the water to the empty end of the box, as shown in *Figure 20*. Begin mixing this into the dry materials. This can be easily done from the water end of the box using the hoe with short pull-and-push strokes. Continue in this fashion with the hoe at a 45-degree angle until all the material is well mixed. Add the remaining water to obtain the desired consistency.

Figure 20. Add Half The Water To The Empty End

Step 5   After the mortar is mixed, pull it to one end of the box to prevent it from drying out, as shown in *Figure 21*. At this time, the mortar can be checked to see if it is of the proper consistency and workability.

Figure 21. Pull The Batch To One End

Step 6   Pick up a small amount of mortar on a trowel and set it firmly on the trowel by tapping it once on the side of the box. Turn the trowel upside down. If the mortar is the proper consistency, it will remain on the trowel as shown in *Figure 22*. This is also a measure of the mortar's adhesion. If the mortar does not adhere, use the corrective techniques discussed in the next section.

Figure 22. Check For Workability And Adhesion

Step 7   Transport the mortar to the work site with a wheelbarrow, as shown in *Figure 23*. Before using the wheelbarrow, wet the inside surface with water. This will prevent the mortar from sticking to the wheelbarrow. When only a small amount of mortar is needed, it can be mixed directly in the wheelbarrow.

105F23.tif

Figure 23. Using A Wheelbarrow To Transport Mortar

Step 8   The final step is to clean the mixing equipment of all mortar as soon as possible after the mixing box or wheelbarrow is no longer needed. A water hose with a spray nozzle and a stiff brush are the best tools to use when cleaning your masonry tools. *Figure 24* shows this cleanup. You can also use a water barrel as a bath to keep the mortar from drying on the mixing tools.

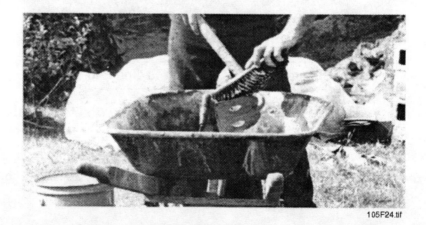

105F24.tif

Figure 24. Clean Up With A Water Hose And Brush

10.0.0   PROBLEMS IN MIXING MORTAR

Typically, the mason faces four general types of problems when mixing mortar: improper proportioning of materials, poor quality of materials, working in cold weather, and retempering. Efflorescence is another mortar problem, although it does not show up until after the mortar has hardened.

10.1.0 PROPORTIONING MATERIALS

Improper proportioning of materials is a common problem in mixing mortar. This is caused primarily by poor work techniques. The first step in preparing the mix is to adequately plan. Ensure that the mix ratios are completely understood and the formula written down. The mason preparing the mix should go through a mental checklist of the things needed and the steps required for the mixing operation. Planning ahead will bring out any missing pieces or shortages that would lead to problems during the mixing.

Here are the four most common problems:

- Adding too much water. This is caused by using a hose to add water. Adding excessive amounts of water is known as drowning the mortar. This lowers the water-cement ratio, which leads to lower mortar strength. This can be corrected by adding additional dry ingredients in their proper proportions until the desired consistency is reached. The best way to avoid adding too much water is to premeasure the water in a bucket. This is important because trying to correct this error can lead to new problems.

- Adding too much sand. This is caused by not measuring your materials. If you use a shovel to initially fill the power mixer or mortar box, calibrate it ahead of time. Do this by filling a cubic foot box or other container of known volume. Once this shovel is calibrated, use it. Also remember that the amount of sand shoveled depends on the moisture content of the sand and how tired you are. You should calibrate the shovel at least twice a day, once before you start in the morning and once midway through the day. If you continue to have problems with measuring the sand, using the cubic foot box may be the best answer. This is actually a faster way to load sand, but you will need a helper to help lift the box and empty it into the drum.

- Adding too much sand to counteract the effects of too much water. This practice results in weaker mortar and should be avoided. Overly sanded mortar is harsh, difficult to use, and forms weak bonds with the masonry units. The color of the mortar will also vary. Never add additional sand to a mix without adding the other ingredients in proportion.

- Adding too much or too little cement. Mortar that is too high in cement content is known as fat mortar. It is sticky and hard to remove from the drum. Mortar that is low in cement content is called lean mortar. It will display all the characteristics of weak mortar, such as poor bonding, lack of cohesion and adhesion, and lack of compressive and tensile strength. Both of these conditions can be avoided by always mixing a one-bag mix. That is, mix a mortar batch of a size that requires one bag each of cement and lime, or one bag of masonry cement, and the correct amount of sand.

All these problems can be avoided by proper planning, thinking ahead to anticipate problems, using calibrated containers, and approaching the mixing process in a methodical manner.

10.2.0 POOR QUALITY MATERIALS

Problems with poor quality or spoiled materials can be solved by properly storing, inspecting, and testing materials prior to use. Do not use cement that has knots or lumps; it will not mix well, and taking the lumps out will waste time.

Lay out the mixing area to the best advantage to prevent good material from being spoiled and to increase the efficiency of the operation. For example, by placing the mixer downhill from the cement pallets, you will avoid getting them wet from possible water spills. By insuring a pathway through the mixing site for the wheelbarrows, you ensure that stockpiles are kept clean. Driving wheelbarrows through the stockpiles while transferring mixed mortar is almost a guarantee of spilling mortar on the dry materials.

10.3.0 COLD WEATHER

Cold weather presents special problems in mixing mortar. Certain admixtures can be added to improve the performance of mortar in cold weather.

Air-entraining admixtures will increase the workability and freeze-thaw durability of mortar, but too much will reduce the strength of the mortar. Adding an air-entraining admixture separately at the mixer is not recommended. This is due to the lack of control over the final air content. If air-entraining is needed, use cements and lime that have been modified at the factory to include an air-entraining agent.

At low temperatures, performance can be improved by using Type III (high early strength) cement. Also, mortar made with lime in dry form is preferred for winter use over slaked quicklime or lime putty because it requires less water.

Another option for cold weather is to heat the material used to make the mortar. The sand and the water can both be heated, though it is easier and safer to heat the water. Do not heat the water over 160°F, however, because of the danger of a flash set when the cement is added to the mix. A flash set occurs when the cement sets prematurely due to excessive heat. Heating of the materials should be considered when the temperature falls below 40°F. After mixing the heated ingredients, the temperature should be between 70°F and 105°F.

There are some admixtures that are not recommended due to their negative effect on mortar. Any antifreeze admixtures, including several types of alcohol, would have to be used in such large quantities that the strength of the mortar would be greatly reduced. Calcium chloride, commonly used as an accelerator in concrete, produces increased shrinkage, efflorescence, and metal corrosion in mortar. Its use is not allowed.

10.4.0 RETEMPERING

Fresh mortar should be prepared at the rate it is used so that its workability will remain about the same throughout the day. Mortar that has been mixed but not used immediately tends to dry out and stiffen. However, loss of water by absorption and evaporation on a dry day can be reduced by wetting the mortar board and by covering the mortar in the mortar box, wheelbarrow, or tub.

If necessary to restore workability, mortar may be retempered by adding water. Thorough remixing is then necessary. Remixing can be done in the wheelbarrow or on a separate surface such as a mortar board. Although small additions of water may slightly reduce the compressive strength of the mortar, the end effect is acceptable. Masonry built using a workable plastic mortar has a better bond strength than masonry built using dry, stiff mortar. Mortar used within one hour after mixing should not need retempering unless the weather is very hot and evaporative conditions prevail.

Mortar that has stiffened because of hydration hardening should be discarded. It is difficult to determine by sight or feel whether mortar stiffening is due to evaporation or hydration. The most practical method of determining the suitability of mortar is on the basis of time elapsed after mixing. Mortar should be used within 24 hours after the initial mixing to avoid hydration hardening.

Colored mortar is very sensitive to retempering; additional water may cause a noticeable lightening of the color. It should be retempered with caution to avoid variations in the color of the hardened mortar. Mixing smaller batches can lessen the need to retemper.

10.5.0 EFFLORESCENCE

Efflorescence is a deposit of water-soluble salts upon the surface of a masonry wall. It is usually white in color and generally appears soon after the wall has been built—when the owner and architect tend to be most concerned with the structure's appearance. Efflorescence occurs when soluble salts and moisture are both present in the wall. If either of those elements are missing or below a certain concentration, efflorescence will not occur. *Figure 25* shows a brick wall of an apartment building which has been discolored by the salt deposit seeping out of the brick.

The salts usually originate in the walls, in either the masonry units or the mortar itself. In some cases, the salts may come from ground moisture behind a basement wall. For this reason, basement walls should be protected by using some sort of moisture barrier.

Figure 25. Efflorescence On A Brick Wall

Efflorescence can be prevented in several ways.

• Reduce salt content in the mortar materials by using washed sand.
• Use clean water for making the mortar.
• Keep materials properly stored and off the ground prior to use.
• Protect newly built masonry walls with a canvas or other suitable waterproofing material.
• Use quality workmanship to insure strong tight bonds between mortar and masonry units.

Efflorescence can be removed by washing the wall with water or a muriatic acid-water solution. Use a diluted solution of one part muriatic acid to nine parts water. This washing will remove the efflorescence, but it may return after a period of time. The best solution is to reduce the amount of moisture entering the wall and to lower the amount of water-soluble salts in the mortar materials.

SUMMARY

This module introduces the trainee to masonry mortar. Topics include the basic ingredients in masonry mortar, properties of different types of mortar, preparations for mixing mortar, techniques for mixing mortar both manually and by machine, admixtures and their characteristics, and some common problems encountered in mixing mortar.

A detailed step-by-step mixing procedure for both manual and machine mixing is presented, as well as safety tips for using power mixers.

References

For advanced study of topics covered in this task module, the following books are suggested:

Building Block Walls—A Basic Guide, National Concrete Masonry Association, Herndon, VA, 1988.

Masonry Design and Detailing—For Architects, Engineers and Contractors, Fourth Edition, Christine Beall, McGraw-Hill, New York, NY, 1997.

Masonry Skills, Third Edition, Richard T. Kreh, Delmar Publishers, Inc., Albany, NY, 1990.

The ABCs of Concrete Masonry Construction, Videotape, 13:34 minutes, Portland Cement Association, Skokie, IL, 1980.

ACKNOWLEDGMENTS

Figures 1-7 and 9 courtesy of Roy Jorgensen Associates, Inc.

Figure 8 courtesy of Associated General Contractors

Figures 10-25 courtesy of Delmar Publishing

REVIEW/PRACTICE QUESTIONS

1. Bond strength refers to _____.
    a. how well the mortar adheres to the masonry unit
    b. the mortar's compressive strength
    c. the mortar's cohesive nature
    d. the mortar's percentage of cement

2. Mortar is commonly referred to in the trade as _____.
    a. concrete
    b. gypsum
    c. cement
    d. mud

3. A mortar mix with insufficient cement produces a weak, harsh mix that lacks strength and adherence. This type of mortar is called _____.
    a. mud
    b. sharp mortar
    c. lean mortar
    d. fat mortar

4. Typically, the time for mixing mortar in a power mixer is _____ after all ingredients are added.
    a. 2 minutes
    b. 10 to 12 minutes
    c. 3 to 5 minutes
    d. 6 to 8 minutes

5. If too much water has been added to the mix, the mason should _____.
    a. add additional sand
    b. add additional cement
    c. pour off the excess water
    d. add all dry ingredients in proportion to the original formula

6. The siltation test for sand is meant to determine _____.
    a. acceptable limits for sodium chloride
    b. how well color will adhere to the sand
    c. the presence of organic and other undesirable elements
    d. the strength of the sand

7. Of the five types of mortar, which is the strongest?

    a. Type K
    b. Type N
    c. Type M
    d. Type O

8. Mixing 1 cubic foot of cement, 1 cubic foot of lime, 6 cubic feet of sand, and enough water to obtain a 5-inch slump will result in approximately _____ cubic feet of mortar.

    a. 6
    b. 8
    c. 2
    d. 12

9. For cold weather masonry work, at what temperature should you begin heating the mortar ingredients?

    a. 32°F
    b. 40°F
    c. 45°F
    d. 50°F

10. Chlorine admixtures are not recommended for mortar because chlorine _____.

    a. is a health hazard to the mason
    b. causes bleaching of the mortar's color
    c. causes efflorescence
    d. causes the mortar to remain plastic too long

11. The _____ of mortar can be tested by placing it in a trowel and turning the trowel upside down.

    a. strength
    b. adhesion
    c. workability
    d. water retention

12. Efflorescence is caused by _____.

    a. too much white lime in the mix
    b. too much water-soluble salt in the mix
    c. hot weather during the construction of the structure
    d. certain types of color admixtures

13. Water retention in mortar is desired to _____.

    a. increase the mortar workability
    b. keep the mortar cool
    c. increase the mortar's ability to retain color
    d. decrease the chance of efflorescence

14. Which type of cement is used most often by masons to mix mortar?
    a. Type I
    b. Type II
    c. Type III
    d. Type IV

15. Mortar gains compressive strength as _____ is added to the mix.
    a. cement
    b. lime
    c. sand
    d. water

16. One of the advantages of using lime in mortar is its characteristic of autogenous healing, or the ability to _____.
    a. reseal the effects of efflorescence
    b. resist the effects of airborne acids and alkalis
    c. reseal small cracks in joints
    d. resist the strength-weakening effects of air-entraining admixtures

17. Retempering mortar is done to extend its workability after the mortar _____.
    a. freezes and then thaws out
    b. was mixed too lean
    c. dries out within the first 1½-hours
    d. proves to be too fat

18. If too much color pigment is added to the mix, the mortar will lose _____.
    a. workability and water retention
    b. strength and durability
    c. water retention and consistency
    d. durability and autogenous healing

19. Water should be placed first into the power mixing drum in order to _____.
    a. cool it off on hot days
    b. wet the inside of it so that the cement will not stick to it
    c. increase the workability of the mix
    d. help speed up the mixing process

20. Adding _____ has no appreciable effect on workability.
    a. water reducers and color
    b. plasticizers and set accelerators
    c. air-entraining and pozzolanic agents
    d. bonding agents and water repellents

ANSWERS TO REVIEW/PRACTICE QUESTIONS

Answer	**Section Reference**
1. a	4.2.3
2. d	7.0.0
3. c	10.1.0
4. c	8.1.0
5. d	10.1.0
6. c	2.4.0
7. c	3.0.0
8. a	7.3.0
9. b	10.3.0
10. c	10.3.0
11. b	9.0.0
12. b	10.5.0
13. a	4.1.2
14. a	2.1.0
15. a	4.2.2
16. c	2.2.0
17. c	10.4.0
18. b	6.0.0
19. b	8.1.0
20. d	5.0.0/Figure 7

The NCCER makes every effort to keep these manuals up-to-date and free of technical errors. We appreciate your help in this process. If you have an idea for improving this manual, or if you find an error, a typographical mistake, or an inaccuracy in the NCCER's Craft Training Manuals, please write us, using this form or a photocopy. Be sure to include the exact module number, page number, a description of the problem, and the correction, if possible. Your input will be brought to the attention of the Technical Review Committee. Thank you for your assistance.

Instructors – If you found that additional materials were necessary in order to teach this module effectively, please let us know so that we may include them in the Equipment/Materials list in the Instructor's Guide.

Write: Curriculum and Revision Department
National Center for Construction Education and Research
P.O. Box 141104
Gainesville, FL 32614-1104
Fax: 352-334-0932

Craft _____ Module Name _____

Module Number _____ Page Number(s) _____

Description of Problem _____

_____

_____

_____

(Optional) Correction of Problem _____

_____

_____

(Optional) Your Name and Address _____

_____

_____

notes

Masonry Units and Installation Techniques

Module 28106

Masonry Trainee Task Module 28106

MASONRY UNITS AND INSTALLATION TECHNIQUES

NATIONAL
CENTER FOR
CONSTRUCTION
EDUCATION AND
RESEARCH

OBJECTIVES

Upon completion of this module, the trainee will be able to:

1. Describe the most common types of masonry units.

2. Describe and demonstrate setting up a wall.

3. Lay a dry bond.

4. Spread and furrow a bed joint and butter masonry units.

5. Describe the different types of masonry bonds.

6. Cut brick and block accurately.

7. Lay masonry units in a true course.

Prerequisites

Successful completion of the following Task Modules is recommended before beginning study of this Task Module: Core Curricula; Masonry Level 1, Modules 28101 through 28105.

Required Trainee Materials

1. Trainee Task Module
2. Pencil and paper
3. Appropriate Personal Protective Equipment
4. Mason's modular rule
5. Framing square
6. 4-foot plumb rule
7. Chalk line
8. Mason's line and trigs
9. Corner blocks or corner poles
10. Marking crayon
11. Trowel
12. Brick hammer
13. Brick set
14. Concave jointer
15. Dusting brush
16. Five-gallon pail

COURSE MAP

This course map shows all of the task modules in the first level of the Masonry curricula. The suggested training order begins at the bottom and proceeds up. Skill levels increase as a trainee advances on the course map. The training order may be adjusted by the local Training Program Sponsor.

LEVEL 1 COMPLETE

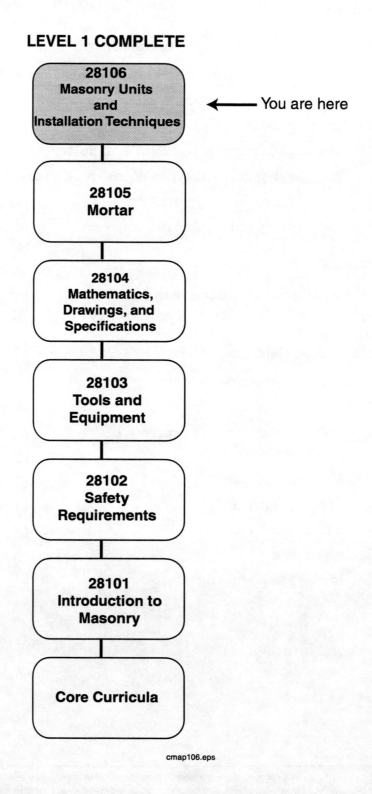

28106
Masonry Units
and
Installation Techniques

← You are here

28105
Mortar

28104
Mathematics,
Drawings, and
Specifications

28103
Tools and
Equipment

28102
Safety
Requirements

28101
Introduction to
Masonry

Core Curricula

cmap106.eps

TABLE OF CONTENTS

TABLE OF CONTENTS (Continued)

Trade Terms Introduced In This Module

Bed joint: A horizontal joint between two masonry units.

Butter: To apply mortar to the ends of a masonry unit.

Closure unit: The last brick or block to fill a course.

CMU: Concrete Masonry Unit.

Course: A row or horizontal layer of masonry units.

Crowding the line: A person touching the mason's line, or a masonry unit too close to the line.

Dry bonding: Laying out masonry units without mortar to establish spacing.

Furrowing: Making an indentation with a trowel point along the center of the mortar bed joint.

Head joint: A vertical joint between two masonry units.

Lead: Part of a wall built as a laying guide and for attaching mason's line

Nominal dimension: The size of the masonry unit plus the thickness of one standard (½ inch or ⅜ inch) mortar joint.

Pointing: Troweling mortar or mortar repairing material, such as epoxy, into a fresh joint after masonry is laid.

Rackback: A lead or other structure built with each course of masonry shorter than the course below it.

Racking: Shortening each course of masonry by one unit so it is shorter than the course below it, resulting in a pyramid shape.

Ranging: Aligning a corner by using a line. Corners can be ranged around themselves or from one corner to another.

Return: A corner in a structure or lead.

Slack to the line: Masonry units set too far away from the mason's line.

Spread: A row of mortar placed into a bed joint.

Stringing: Spreading mortar with a trowel on a wall or footing for a bed joint.

Tail: To check the spacing of head joints by checking the diagonal edges of the courses on a lead or corner.

Tuckpointing: Replacing hardened mortar in joints with new mortar.

1.0.0 INTRODUCTION

This module contains detailed information on masonry materials. It also provides step-by-step instructions for building a single-wythe masonry wall. This module gives you details about the following topics:

- Cement, clay, and stone masonry units
- Setting up and laying out a wall
- Spreading mortar
- Bonding masonry units
- Cutting masonry units
- Laying masonry units
- Mortar and other joints
- Patching, **pointing,** and **tuckpointing**
- Cleaning masonry units

1.1.0 SAFETY

Safety is a continuous effort. Being a mason calls for following safe work practices and procedures. This includes the following work activities:

- Inspect tools and equipment before use.
- Use tools and equipment properly.
- Keep tools and equipment clean and properly maintained.
- Keep your hands out of the mortar.
- Remember to use the right tool for the job.
- Assemble scaffolding and foot boards properly.
- Use caution and common sense when working on elevated surfaces.

Accidents also happen when tools and equipment are left in the way of other workers. Store tools safely, where other people cannot trip over them and where the tools cannot get damaged. Clean, well-kept tools make for safe work as well as good work.

Do not drop or temporarily store tools or masonry units in pathways or around other workers. Stack masonry units neatly so the stack will be less likely to topple over. Stack materials by reversing the direction of the units on every other layer so they will be less prone to tip. Keep the pile neat and vertical to avoid snagging clothes. Do not stack masonry units higher than your chest (as a rule). A stack which is too high is more likely to tip over.

Various materials used to build masonry structures have properties that can be harmful to the mason. Mortar and grout, for example, can be very caustic. This means that continued contact with the skin can cause burns and irritation. It is important to protect your skin and eyes when mixing and working with these materials.

Cutting masonry units can be a dangerous job in several different ways. When sawing or chiseling a masonry unit, chips can fly off and hit your eyes.

CAUTION: Always wear safety glasses or goggles and respiratory protection when cutting masonry units.

The process of using a saw or chisel is dangerous. Always keep your hands away from the blade. Do not operate power saws when you are tired or sick.

2.0.0   CONCRETE MASONRY MATERIALS

Most masonry materials are made of clay, concrete, or stone. The masonry unit most commonly used in the United States is the concrete block. The term **CMU** stands for concrete masonry unit. CMUs are classified into six types:

- Loadbearing concrete block
- Nonloadbearing concrete block
- Concrete brick
- Calcium silicate units
- Prefaced or prefinished concrete facing units
- Concrete units for manholes and catch basins

2.1.0   ASTM SPECIFICATIONS

Each of the above types of unit has its own American Society for Testing and Materials (ASTM) standards for performance characteristics. The standards describe the expected performance of the CMU in compressive strength, water absorption, loadbearing, and other characteristics. CMUs are valuable for their fire resistance, sound absorption, and insulation value.

2.1.1   Compressive Strength

The compressive strength of a CMU measures how much weight it can support without collapsing. These figures are set according to ASTM test results. The tests are performed on a specific shape, size, and weight of CMU. The CMU tested then becomes the standard for that particular compressive strength. Compressive strength is measured in pounds of pressure per square inch, or psi. This is the weight that the unit, and the structure made of the units, can support. A structure supports less imposed weight than the sum of its unit strength because the structure has to support itself as well.

The work of the mason is an important part of how a masonry structure performs and whether it meets its compressive strength specification. Tests have shown that the compressive strength of a loaded wall is about 42 percent of the compressive strength of a single CMU when the mason uses a face shell mortar bedding. When the mason uses a full mortar bedding, the compressive strength increases to about 53 percent. The engineer factors these components into the equation for picking the CMU. The factoring is contained in the job specifications. This is another reason that specifications are important.

2.1.2 Moisture Absorption And Content

Moisture in the CMU has an effect on shrinking and cracking in the finished structure. Generally, the lower the moisture, the less likely the units are to shrink after they are set. Acceptable moisture content and absorption rates are set by ASTM standards and local codes. Manufacturers specify that their units meet ASTM or other standards. Unit tests are made to make sure they do.

For the mason, this means keeping CMUs dry. Never wet CMUs immediately before or during the time they are to be laid. Stockpile them on planks or pallets off the ground. Use plastic or tarpaulin covers for protection against rain and snow, as shown in *Figure 1*.

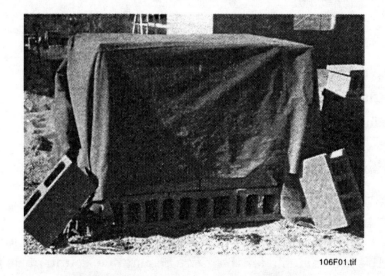

106F01.tif

Figure 1. Protecting Masonry Materials

When stopping work, cover the tops of masonry structures to keep rain or snow off. Be sure moisture does not get into cavities between wythes. When laying CMUs for interior use, dry them before laying. They should be dried to the average condition to which the finished wall will be exposed.

2.2.0 CONTRACTION AND EXPANSION JOINTS

CMUs have one major difference from other masonry units. CMU construction is prone to shrinkage cracking as are concrete slabs and sidewalks. Shrinkage cracking occurs as the concrete slowly finishes drying. As the concrete shrinks, it moves slightly. The movement causes cracks in a rigid slab or wall. The shrinkage cracking in a CMU structure is controlled in the same way as for concrete slabs. This is done by inserting reinforcement in combination with cutting some contraction or control joints.

Figure 2 shows some typical locations for contraction joints. Walls are likely to crack at abrupt changes in wall thickness or heights, at openings, over windows, and over doors. Contraction joints are also used at intersections between loadbearing walls and partition walls.

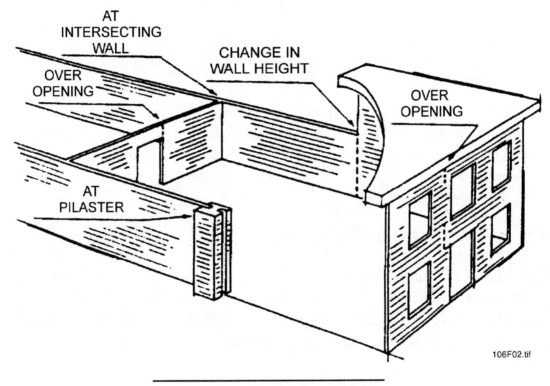

AT INTERSECTING WALL

OVER OPENING

CHANGE IN WALL HEIGHT

OVER OPENING

AT PILASTER

106F02.tif

Figure 2. Location Of Control Joints

CMU walls use two kinds of reinforcement: grout, and steel with grout. The grout is poured into CMU cores, or steel rods are inserted in CMU cores and grout is poured around the steel. The reinforcement gives rigidity and strength to the wall. CMU walls need reinforcement of either kind on both sides of a contraction joint.

Contraction joints control cracking by weakening the structure. Cracks occur at the weakened area instead of randomly. In a concrete slab, these control joints are made by cutting grooves. In a CMU wall, this is done by breaking the contact between two columns of units, as shown in *Figure 3*.

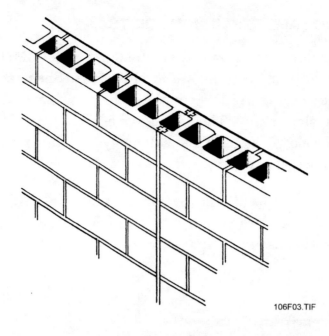

Figure 3. Vertical Control Joint With Plastic Filler

The control joints replace a standard mortar joint every 20 feet or so. Control joints must be no more than 30 feet apart. The control joints in CMU walls have no mortar; they are filled with silicon or another material that is not rigid. The contraction joint filler keeps the rain out but allows slight movement. The reinforcement keeps the edges of the contraction joints aligned as they move.

Clay masonry units do not contract and shrink as they age. But they do change size very slightly with temperature and moisture changes in the air. Clay masonry structures need expansion joints to handle this type of movement. The control joints in CMU walls take care of expansion as well as contraction. Clay masonry walls must include expansion joints. These are usually soft, mortarless joints filled with foam and covered with a layer of silicon paste. As with contraction joints, the filler keeps the rain out and allows slight movement of the masonry.

2.3.0   CMU MANUFACTURING

Concrete blocks are made of water added to portland cement, aggregates, and admixtures. Cement is one of the few chemical compounds that harden in water. It is very versatile. The major difference between CMUs, mortar, grout, and concrete slab floors lies in the aggregates, the admixtures, and the proportions of the concrete mix.

* Portland cement is a mixture of lime plus silica, iron, and alumina ores from limestone, shale, and clay. They are proportioned by weight according to the type of cement being made. The dry ingredients get ground together and burnt. The residue of clinkers is reground to a fine powder that is the cement.

MASONRY — TRAINEE TASK MODULE 28106

- Aggregates are materials that give concrete texture and weight. Common aggregates are sand, gravel, crushed limestone, and finely ground slag. Lightweight aggregates include coal cinders, fly ash, pumice, scoria, and specially treated shale, clay, or slag. The aggregates give the concrete strength and weight.

- Admixtures affect the properties of the finished concrete. Common admixtures add color, entrain air in the mixture, slow setting time, repel moisture, or hold extra water. Admixtures affect properties such as freeze resistance, weight, and speed of setting.

Figure 4 shows a typical CMU manufacturing operation. Cement and different aggregates arrive by boxcar and are loaded into hoppers. An automated mixer measures out the dry ingredients by weight and adds water. The concrete paste is churned and mixed. Then it is fed to the block machine where it is dropped into molds. The paste gets compacted and vibrated in the molds to drive out air bubbles and to fill the spaces completely. It takes about six seconds to form a block. The blocks are put on pallets and moved into curing rooms.

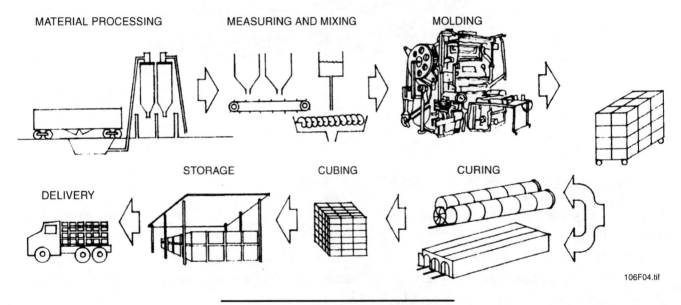

106F04.tif

Figure 4. Modern CMU Manufacturing

The paste hardens through a chemical process called hydration which depends on available moisture. Concrete hardens slowly, but this process can be speeded up. Concrete blocks can be cured in a steam kiln like the one shown in *Figure 5*. The block is racked, then the racks are collected and rolled into the kiln enclosure. The blocks are steamed in the kiln, then take 24 to 28 days to dry depending on their shape.

Blocks can be cured faster with live steam under pressure in an autoclave. The steam provides the heat and moisture needed to speed the curing of the blocks. High-pressure steam curing time averages about 24 hours depending on the process used.

After curing, the blocks are cubed and stored. They finish cooling and drying as they wait to be loaded on trucks for delivery. The process is almost completely automated.

106F05.tif

Figure 5. Blocks Leaving Kiln

2.4.0   CONCRETE BLOCK CHARACTERISTICS

Block is produced in four classes: solid loadbearing and non-loadbearing, and hollow loadbearing and non-loadbearing. Block comes in two weights: normal and lightweight. Lightweight block is made with fly ash, pumice, and scoria or other lightweight aggregate. Loadbearing and appearance qualities of the two weights are similar. The major difference is that lightweight block is easier and faster to lay.

Block is classified as hollow or solid. Solid units are for special needs, such as structures with unusually high loads, drainage catch basins, manholes, or firewalls. As with clay products, solid units have 25 percent or less of their surface hollow. A hollow unit has less than 75 percent of its surface area solid. *Figure 6* shows the names of different parts of blocks.

Most block falls under specification ASTM C 90 which covers hollow loadbearing block. ASTM C 90 specified grades N and S until recently. Grade S has been discontinued for this block, and ASTM C 90 block is now ungraded. The ASTM does specify a minimum compressive strength of 800 psi. Solid loadbearing block falls under ASTM C 145 which specifies grades N and S. Grade N has a compressive strength of 1500 psi and the compressive strength of Grade S is 1000 psi.

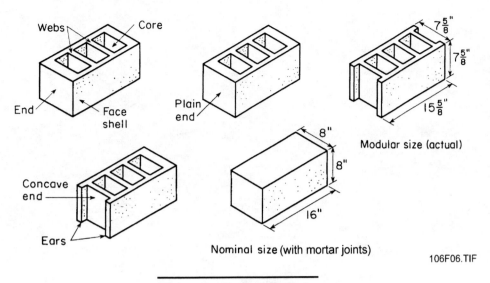

Figure 6. Parts Of Blocks

Block comes in modular sizes, with colors determined by the cement ingredients, the aggregates, or any additives. The basic block is called a stretcher. It has a nominal face size of 8 × 16 inches with a standard ⅜-inch mortar joint. The most commonly used block has a nominal width of 8 inches, but 4", 6", 10", and 12" widths are also common. Three modular bricks have the same nominal height as one nominal 8" block. Two nominal brick lengths equal one nominal block length. So, laying one stretcher covers the same area as laying six modular bricks.

Block comes in a variety of shapes to fit common and special purposes. *Figure 7* shows a small part of the range of block sizes and shapes. Common hollow units can have two or three cores. Most cores are tapered slightly to provide a larger **bed joint** surface. Block edges may be flanged, notched, or smooth. There are local variations as well, with some shapes available only in specific parts of the country.

A variety of surface and mixing treatments can give block varied and attractive surfaces. Newer finishing techniques give block face the appearance of brick, stone, ribbed columns, raised patterns, or architectural fabrics. Just as with clay masonry units, block can be laid in structural pattern bonds.

Loadbearing block is used as backing for veneer walls, for bearing walls, and for all structural assemblies. Both regular and specially-shaped blocks are used for paving, retaining walls, and slope protection. Landscape architects call the newer, shaped-to-interlock blocks "hardscape" and use them as part of landscape design.

Nonstructural block is specified under ASTM C 129, listing a minimum compressive strength of 500 psi. This block is used for screening and non-loadbearing partition walls. Elegantly surfaced, solid nonstructural block is often used as a veneer wall for wood, steel, or other backing. Hollow nonstructural block is made with pattern cores much like clay tile. Pattern core blocks come in a variety of shapes and modular sizes and are commonly used for screen walls.

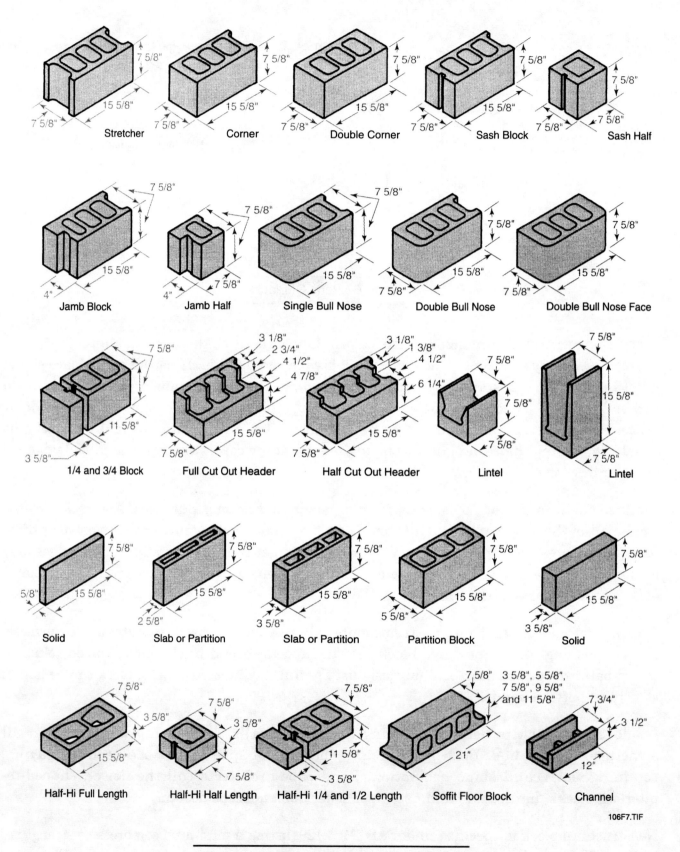

Figure 7. Common Concrete Block Shapes

106F7.TIF

2.5.0 CONCRETE BRICK

Concrete brick is a solid loadbearing unit, roughly brick size, used in the same way as clay brick. Concrete brick has no voids and may be frogged as shown in *Figure 8*. A frog is a depression in the head of a brick that lightens the weight of the brick. It also makes for a better mortar joint by increasing the area of mortar contact.

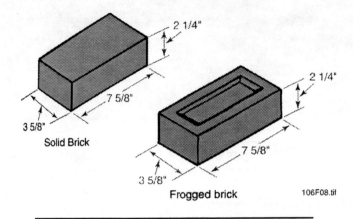

Figure 8. Concrete Brick, Solid And Frogged

Concrete brick is designed to be laid with a ⅜-inch mortar joint. It comes in many sizes, with the most popular **nominal dimensions** of 4 × 8 inches. This size gives 3 courses in a height of 8 inches, just as with standard modular brick.

ASTM C 55 specifies two grades of concrete brick.

* Grade N is used for architectural veneers and facing units in exterior walls. It has high resistance to moisture and frost penetration. It has a compressive strength of 3000 psi.

* Grade S is also used for architectural veneers and facing units. It has moderate resistance to moisture and frost. It is used in the southern region of the United States. Its compressive strength rating is 2000 psi.

Slump brick is made from a wet mixture. The units sag or slump when removed from the molds, as shown in *Figure 9*. This gives an irregular face resembling stone. In other respects, slump brick meets concrete brick standards.

Solid block is also made from a slump mixture. Because of the greater surface area, the block face is very irregular. Its height, surface texture, and appearance resemble stone.

106F09.tif

Figure 9. Slump Brick In A Wall

2.6.0 OTHER CONCRETE UNITS

CMUs include prefaced units, calcium silicate CMUs, and catch basin units. ASTM specifications cover loadbearing, moisture retention, aggregate mix, and other characteristics of these units as well.

2.6.1 Prefaced Units

Concrete prefaced or precoated units are faced with colors, patterns, and textures on one or two face shells. The facings are made of resins, portland cement, ceramic glazes, porcelainized glazes, or mineral glazes. The slick facing is easily cleaned. These units are popular for use in gyms, hospital or school halls, swimming pools, and food processing plants. They come in a variety of sizes and special-purpose shapes, such as coving and bullnose corners. *Figure 10* shows commonly used concrete prefaced units.

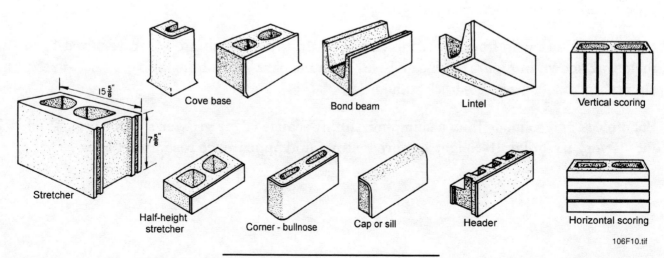

106F10.tif

Figure 10. Prefaced Concrete Units

2.6.2 Calcium Silicate Units

Calcium silicate units are made of a mixture of sand, water, lime, and calcium silicate. The calcium silicate acts as a leavening agent and creates gas bubbles in the mix. The units are not fired or cured in a kiln but cured in an autoclave with pressurized live steam. In the autoclave, the lime reacts with the silica to bind the sand particles into a very lightweight, strong unit. ASTM performance specifications cover this type of brick and block with grading standards identical to those for traditional products.

The units are also called "sand-lime brick" or "aerated block." They are used extensively in Europe, Australia, Mexico, and the Middle East. In the U.S., this brick is used mostly in flues, chimney stacks, and other high-temperature locations. It resists sulfates in soil, does not effloresce, and is not damaged by repeated freeze-thaw cycles. The block is now manufactured in the United States in a variety of sizes for commercial or home building.

2.6.3 Catch Basins

Concrete manholes and catch basin units are specially made with high strength aggregates. They must resist the internal pressure generated by the liquid in the completed compartment. *Figure 11* shows the shaped units manufactured for the top of a catchment vault. These blocks are engineered to fit the vault shape and cast to specification. They are made with interlocking ends for further strength.

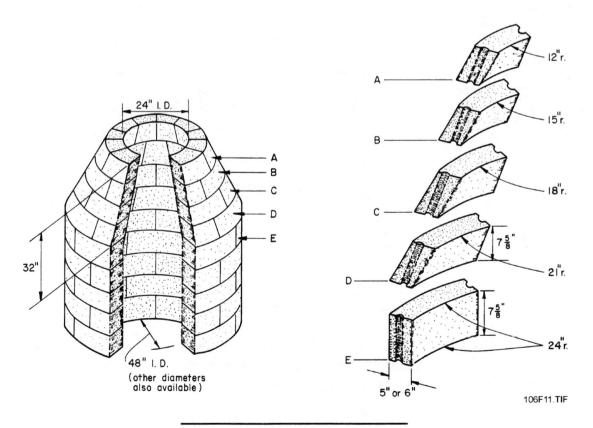

Figure 11. Manhole And Vault Units

3.0.0 CLAY AND OTHER MASONRY MATERIALS

As you learned in *Introduction to Masonry*, clay masonry materials are the second-oldest building material. The following sections review clay and stone masonry units and introduce metal and plastic masonry materials.

3.1.0 CLAY MASONRY UNITS

Structural clay masonry units include :

- Solid masonry units or brick
- Hollow masonry units
- Architectural terra cotta units

Solid masonry units have 25 percent or less of their surface as a void or hole. Hollow masonry units have 75 percent or less of their surface as a void and are usually called tiles. ASTM standards cover all types of masonry units, loadbearing and non-loadbearing. Masonry units come in standard modular and nonmodular sizes, in a wide range of colors, textures, and finishes. This module will focus on brick and laying brick.

Bricks have six sides and each side has a name. The name reflects how the brick is used in patterning courses. *Figure 12* shows each of the different faces of brick and their names. The shaded part of the brick is the named part and the part that shows when the brick is laid in a pattern bond.

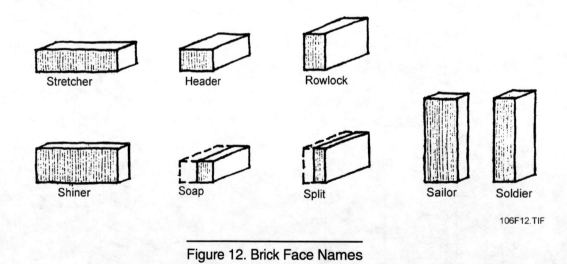

106F12.TIF

Figure 12. Brick Face Names

The most important characteristic of brick is its absorption capacity, or the amount of water it can soak up in a fixed length of time. The percentage of water present in brick affects the hardening of the mortar around the brick. If the brick contains a high percentage of moisture, the mortar will set more slowly than usual and the bond will be poor. The brick will not absorb moisture and mortar into its microscopic irregularities. If the brick contains a low percentage of moisture, it will absorb too much moisture from the mortar. This will prevent the mortar from hardening properly because there will not be enough water left for good hydration.

Hard-surfaced bricks and CMUs usually need to be covered on the job site so they do not get wet. Soft-surfaced bricks are usually very absorbent and may sometimes need to be wetted down before they are used.

The mason needs to determine whether the brick is too dry for a good bond with the mortar. The following test can be used to measure the absorption rate of brick:

Step 1   Draw a circle 1 inch in diameter (around a quarter) on the surface of the brick with a crayon or wax marker.

Step 2   With a medicine dropper, place 20 drops of water inside the circle.

Step 3   Using a watch with a second hand, note the time required for the water to be absorbed.

If the time for absorption exceeds 1½ minutes, the brick does not need to be wetted. If the brick absorbs the water in less than 1½ minutes, the brick should be wetted.

Wet brick with a hose played on the brick pile until water runs from all sides. Let the surface of the bricks dry before laying them in the wall.

3.2.0   STONE

As noted in the first module, rubble stone is irregular in size and shape. Ashlar stone has been cut into a rectangular unit. Stone is expensive to assemble and time-intensive to lay. It is rarely used today except for trim and detail.

Natural stone is used mostly for veneer walls, floors, and trim. Rubble and ashlar are used for dry stone walls, mortared stone walls, retaining walls, facing walls, slope protection, paving, fireplaces, patios, and walkways. Limestone ashlars still make the finest sill blocks. As they are one piece, there is no concern about water coming through into the wall underneath. They are also still used for lintels and as coping stones on top of brick walls.

Concrete masonry units are made in shapes and colorings to mimic every kind of ashlar. These units are called cast stone and are more regular in shape and finish than natural stone. They arc lighter in weight and do not need as large a footing as natural stone. Cast stone has replaced natural stone in most commercial projects because of its economy. ASTM specifications cover cast and natural stone.

Masons may think stone work is a separate craft, but that is not entirely so. Masons still must know how to lay stone copings or sills or lintels. Masons must pattern, shape, and lay stone veneer for fireplaces or home walls.

3.3.0 OTHER MASONRY MATERIALS

Masonry walls usually need material in addition to mortar to make the wall stronger, to hold it in place, or to handle moisture. The masonry contractor or general contractor will buy these materials and supply them as part of the job. Plans and specifications typically detail the locations and types of these materials to be used.

3.3.1 Metal Ties

Metal ties are used to tie together cavity walls. The ties keep the walls from separating when weight is placed on them by the other parts of the structure. Metal ties are also used for composite walls. The ties equalize the loadbearing and also tie the two wythes together. Ties are made from $\frac{3}{16}$-inch zinc coated steel, and are placed 24 inches apart from each other. *Figure 13* shows rectangular and z-shaped ties.

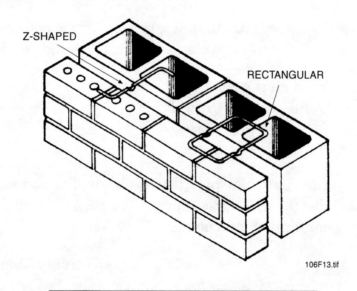

106F13.tif

Figure 13. Rectangular And Z-Shaped Ties

3.3.2 Veneer Ties

Veneer ties are used to tie masonry veneer walls to a backing wall or wythe. Unlike metal ties, veneer ties do not equalize loadbearing. They do keep the veneer wall from moving away from its backing. They are made of corrugated galvanized steel and placed about 12 inches apart. *Figure 14* shows veneer ties.

106F14.tif

Figure 14. Corrugated Veneer Ties

3.3.3 Reinforcement Bars

Steel reinforcement bars come in different thicknesses and lengths. They are inserted in block cores and then the cores are filled with grout. They add strength and weight-bearing capacity to block walls. Sometimes they are placed in the middle of cavity walls where the cavity is to be grouted.

3.3.4 Joint Reinforcement Ties

Joint reinforcement ties are made of two 10-foot lengths of steel bars welded together by rectangular or triangular cross bracing. *Figure 15* shows the ladder (rectangular) and truss (triangular) versions of these ties. They are used in horizontal joints every second or third **course** as specified.

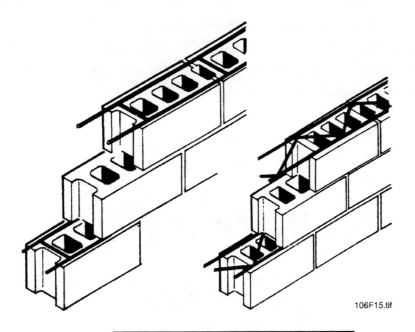

106F15.tif

Figure 15. Steel Joint Reinforcements

3.3.5 Flashing

Flashing keeps water from leaking from the top of a masonry wall into the unit below. It is placed under masonry lintels, sills, copings, and spandrels. It is not needed under stone lintels, sills, and copings. The most common flashing is made of copper, stainless, or galvanized metal. Bituminous flashing is made of fabric saturated with asphalt. Newer types of flashing are made of plastics. They are cheaper and easier to work with. *Figure 16* shows flashing in position under a sill and a lintel.

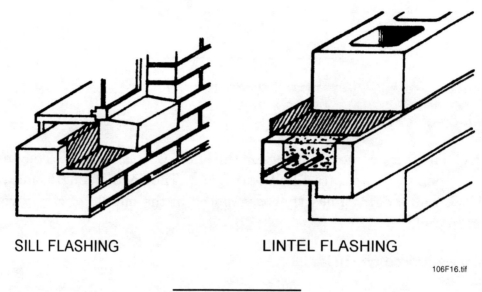

SILL FLASHING   LINTEL FLASHING

106F16.tif

Figure 16. Flashing

3.3.6 Joint Fillers

Plastic or rubber joint fillers are used to replace mortar in expansion or contraction joints. They break the bond between adjacent masonry units and allow expansion and contraction of the wall. They fill the control or expansion joints in order to keep moisture out of the space. *Figure 17* shows molded joint fillers for CMUs.

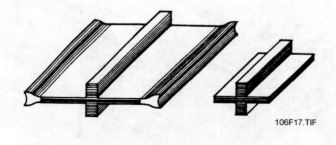

106F17.TIF

Figure 17. Joint Fillers

3.3.7 Anchors

Different kinds of metal bars, bolts, straps, and shaped ties are used to anchor a wall that meets another wall at a 90-degree angle. They are also used to tie different architectural elements to masonry walls. Anchors must be installed according to the specifications, as they affect the loadbearing of the wall. *Figure 18* shows several types of shaped anchors.

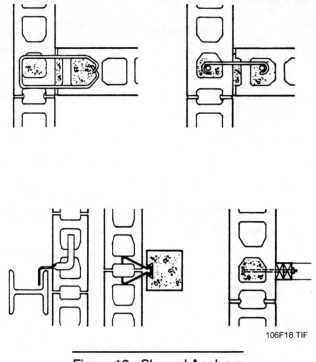

106F18.TIF

Figure 18. Shaped Anchors

4.0.0 SETTING UP AND LAYING OUT

Setting the job up and laying the structure out are two distinct steps. Both must be complete before the mason can start to lay units. Setting up refers to materials and site preparation. Laying out refers to establishing the baseline for the masonry structure. The next sections give details for both of these tasks.

4.1.0 SETTING UP

Setting up masonry work starts when the contract for the job is signed. The first step for the masonry contractor is to read the contract, blueprints, and specifications. The next step is to review the schedule plus any standards and codes cited in the contract. After all of that, the masonry contractor is ready to estimate the workers, materials, and equipment needed for that job. Review the information in *Mathematics, Drawings, and Specifications* to get a clearer idea of what this work entails.

The next step is to estimate again, check figures, and order the masonry equipment and materials. A visit to the job site and discussion with the engineer or construction foreman will give the masonry contractor an idea of where and how to store masonry materials. The masonry contractor must specify a delivery date and location on site. Materials must be stored close to where they will be used and protected from the weather. The crew must be hired and briefed. Then, the work is ready to begin, but there is still much to do before laying the first masonry unit. The following checklist shows some of the preliminary procedures:

- Check that all materials are stored close to work stations and protected from moisture. Masonry units must be laid dry in order to avoid shrinkage upon drying. Pile materials on pallets or planks off the ground and cover with a tarpaulin or plastic sheet. Bagged materials can be stored in sheds or stacked on pallets and covered. Sand must be covered also, to protect it from moisture and dirt.

- Prepare mortar mixing areas within several feet of the work areas or as close as possible. Place water barrels next to them for water supply and for storage of hoes and shovels not in use. Be sure mixing equipment does not interfere with movement paths.

- Place mortar pans and boards by workstations. If scaffolding is used, place them at intervals on the scaffolding near the point of final use.

- Stockpile units on each side of the mortar pans and at intervals along the wall line. If scaffolding is used, place units along the top of the scaffold near the point of final use. Stockpiles should allow the mason to move block as little as possible once laying starts.

- Stack block in stockpiles with the bottom side down, just as they will be laid in the wall. The top of the block has a larger shell and web, as shown in *Figure 19*. Stack faced units with the faced sides in the direction they will go, just as they will be laid in the wall. Stack all units so the mason will move or turn them as little as possible.

- Check all scaffolding for proper assembly and position. Braces should be attached and planks secured at each end. Scaffolding should be level and no closer than 3 inches from the wall.

- Check all mechanical equipment, power tools, and hand tools. Make sure they are clean, in good condition, and the right size for the job.

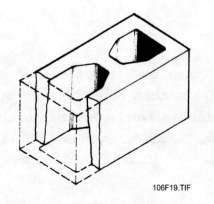

106F19.TIF

Figure 19. Larger Shell On Top Of Block

The contractor may assign a helper to keep mortar pans full by supplying mortar from the mixer. A helper may also be assigned to keep the stockpiles refilled. The objective of all setup work is to make everything efficient and convenient for the masons once they begin laying.

4.2.0 LAYING OUT

Laying out the wall or other structural unit calls for a review of the plans and specifications. The first steps are to plan out the work, establish where it will go, and then to lay a dry bond.

4.2.1 Planning

Planning out the work calls for checking the plans for wall lengths, heights, door dimensions, and window openings. What pattern or bond is specified? What is the nominal size of the masonry unit? How are openings to be treated? Are the dimensions and the masonry units on the modular scale of 4-inch increments?

After answering these questions, the mason can draw a rough layout of the wall and lay out the bond pattern. If the job is sized on the modular grid, graph paper might be handy for the spacing drawing. This drawing can show where the bond pattern will start and how it will fit around the specified openings. From this drawing, the mason can count and calculate how many masonry units to cut.

The question of whether the designer used or did not use the modular grid becomes important. *Figure 20* shows door and window openings located in a running bond.

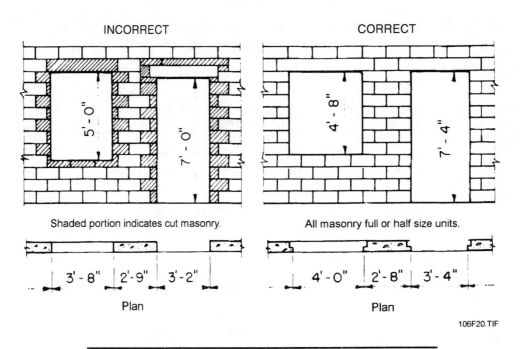

106F20.TIF

Figure 20. Nonmodular And Modular Opening Locations

Notice how the openings are set off the modular grid in the diagram on the left. The amount of cutting is enormous compared to the example on the right. Using many small units reduces wall strength as well. Sometimes the mason can persuade the designer or engineer to shift the openings slightly to avoid so much cutting. In other cases, the dimensions are critical and cannot be changed.

Drawing the bond pattern on the wall area may seem like a time-consuming exercise, but it can save much time, especially with nonmodular work. The starting point of the pattern determines how many masonry units will need to be cut. By adjusting the starting unit of a nonmodular bond pattern, the mason can come up with a layout that calls for cutting the smallest amount of units. The mason will check these calculations by laying a dry bond before cutting any units.

4.2.2 Locating

The mason will check the location first. Masonry walls take a footing or support, usually made of concrete. The surveyor or foreman will mark the corners of the structure or slab. On some jobs, the foreman will drive nails into the wall footing to mark the building line. *Figure 21* shows a foundation layout with a footing plan for block foundation walls. Notice the dimensions for the location of the walls are given in relation to the footing. How can you calculate the distance from the edge of the footing to the beginning of the block wall?

At the job site, the first thing to do is to locate the footing. Next, brush it off. Remove any dried concrete particles or large aggregates to ensure a good bond between the footing and the first course.

Check that the footing is level. If the footing is not within an inch of level, it must be fixed. Do not apply a thick mortar joint to level the first course. This can result in a joint too thick to carry the load of the wall. If the footing is out of level, notify your supervisor.

The next step is to locate the walls. Take measurements from the foundation or floor plan and transfer them to the foundation, footing, or floor slab. All measurements on the plans must be followed accurately. Be sure the door openings are placed exactly and the corners are on the footings exactly as given on the detailed drawings. Check to see that you are not confusing the measurements for the interior and exterior walls. If it appears that the wall cannot be laid out exactly because of errors in the footing, notify your supervisor.

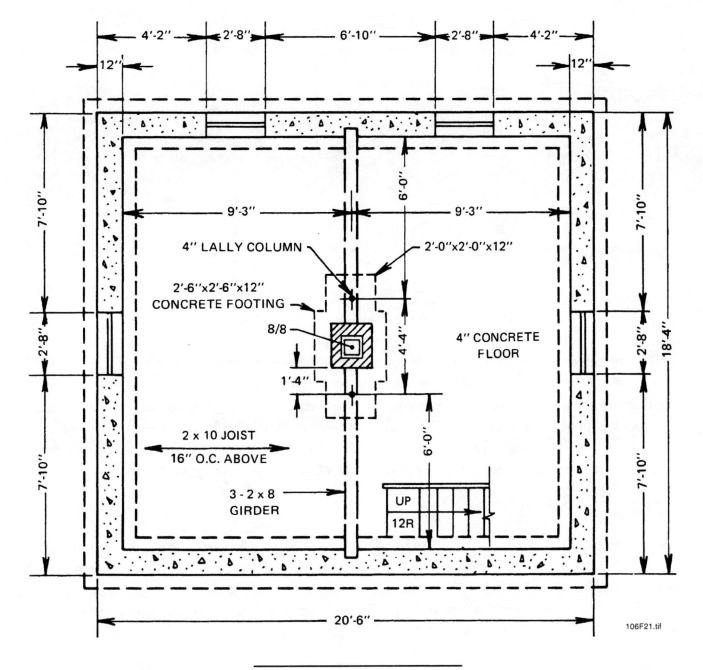

Figure 21. Concrete Footing Plan

The next task is to establish two points, corner-to-corner or corner-to-door. Then, run a chalk line between the two points and snap it on the footing or foundation as in *Figure 22*. Since a chalk line is easily erased, mark key points along the chalk line with a marking pencil, nail, or screwdriver. This will allow resnapping the chalk line without refinding the points.

Figure 22. Snapping A Chalk Line

Mark the entire foundation for walls, openings, and control joints. After snapping the chalk line, mark over the chalk with a marker or nail. After completing all markings, check the measurements of the markings against the foundation plan. Again, be sure you are reading the correct measurements. If there is to be a veneer wall, check that you are dimensioning the veneer, not the backing wythe. If everything does not fit precisely and exactly, it must be done over. It is easier to redo measurements than to redo a masonry wall.

4.2.3   Dry Bonding

As an alternative to using measurement for positioning the masonry units, **dry bonding** can be used. Starting with the corners, the mason can lay the first course with no mortar, or dry bond. This is a visual check of how the units will fit. It also checks the pattern bond drawing and the calculations for cut units. For CMUs, it provides a chance to check unit size and specifications.

From the corners, the mason lays units along the wall markings for the entire foundation, as in *Figure 23*. Since all mortar joints will be standard sizes, use a ⅜- or ½-inch piece of plywood or other material as a spacing jig. Check the specifications for the size of brick joints. Remember that all block is laid with a ⅜-inch joint. If you run into spacing problems, use a spacing jig and mark any adjustments on the foundation and on the jig.

Figure 23. Laying A Dry Bond

Lay the units through door openings to see how bond will be maintained above the doors. Then check spacing for openings above the first course, such as windows. Do this by taking away units from the first course and checking the spacing for the units at the higher level. These checks will show whether the joint width will work out for each course up to the top of the wall. Use the pattern bond diagram to help you. If spacing has to be adjusted slightly, mark it on the diagram and on the foundation.

After the units have been laid out correctly, mark the end of every other unit. Do this with a marking pencil directly on the foundation. This will guide you in laying mortar when the dry units are removed.

Once all of this has been done, the mason can use the steel square to mark the exact location and angle of the corners. The next step is checking the corner layout on the drawings.

The layout of the corner itself is important, especially when you are working with block and modular spacing. The architect will detail the corner layout on the working drawings. Different block layouts, as shown in *Figure 24*, are possible. Each layout takes up a slightly different amount of space. This will affect the modular spacing and whether any block will have to be cut. Building the corners as specified is the key to maintaining modular dimensions.

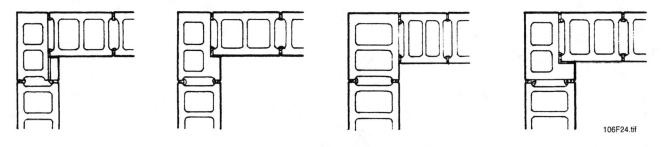

Figure 24. Block Corner Layouts

5.0.0 SPREADING MORTAR

After the mortar is mixed, pick it up on your trowel and spread it. Filling and emptying the trowel is an important part of the mason's skill. Applying the mortar or spreading it is the next step. An apprentice mason must learn to spread mortar so that:

• the joints are completely filled with no small voids for water to enter

• the mortar is still pliable while you level and plumb the unit

• the finished joint is the specified thickness after you level and plumb the masonry unit

• the mortar does not smear the face of the masonry unit

The following sections describe holding the trowel, picking up the mortar, and laying it down. At this point, you will learn something to be experienced rather than memorized. The techniques in the next sections should be practiced until you feel comfortable using them.

5.1.0 PICKING UP MORTAR

There are several ways of using the trowel to pick up mortar. This section will introduce you to a general method for picking up mortar from a board and a general method for picking up mortar from a pan. There are many different ways to do these tasks. The instruction here begins with some tips on holding the trowel. All of these instructions are only approximations of the work itself. You can only learn this through watching a skilled mason and practicing until you feel comfortable with these movements.

5.1.1 Holding The Trowel

Pick up your trowel by the handle. Put your thumb along the top of the handle with the tip on the handle, not the shank, as shown in *Figure 25*.

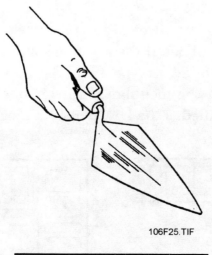

106F25.TIF

Figure 25   Holding The Trowel

CAUTION: Keep your thumb off the shank to keep it out of the mortar. Mortar is caustic and can cause chemical burns.

Keep your second, third, and fourth fingers wrapped around the handle of the trowel. Keep the muscles of your wrist, arm, and shoulder relaxed so you can move the trowel freely.

Most of your work with the trowel will require holding the blade flat, parallel to the ground, or rotating the blade so it is perpendicular to the ground.

Rotating the blade gives you a cutting edge. The best edge for cutting is the edge on the side closest to your thumb. It is best this way because you can see what you are cutting. When you turn the trowel edge to cut, rotate your arm so your thumb moves down. This will rotate the trowel so that the bottom of the blade turns away from you. If you rotate only your wrist, after a while you will strain it. Use the larger muscles in your arm and shoulder to rotate the trowel.

Rotating the blade also gives you a scooping motion. Turning your thumb down will give you a forehand scoop. Turning your thumb up will rotate the bottom of the blade toward you and give you a backhand scoop. Using a forehand or backhand movement will depend on the position of the material you are trying to scoop.

5.1.2   Picking Mortar From A Board

After putting the mortar on the board, follow these steps:

Step 1   Work the mortar into a pile in the center of the board and smooth it off with a backhand stroke.

Step 2   Rotate the trowel edge down. Use the trowel edge to cut off a slice of mortar from the edge. Cut off enough to cover the blade of the trowel.

Step 3   Using the edge of the trowel, pull the mortar a few inches away from the main pile.

Step 4   Pull and roll the slice of mortar to the edge of the board. Work the mortar into a long, tapered roll, as in *Figure 26*.

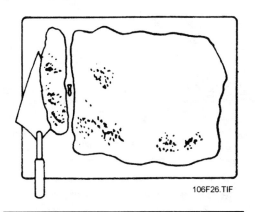

106F26.TIF

Figure 26. Shape The Mortar Into A Roll

Step 5   Hold the trowel blade flat against the mortar board. Slide it under the mortar by pushing it forward. Slide it until the trowel blade is completely under the roll.

Step 6   Raise the trowel blade and the roll off the board by using a very small snap of your wrist. This needs to be a very small snap or you will snap the mortar off the trowel. If that happens, start over.

Raising the trowel quickly will break the bond between the mortar and the board. If done correctly, the mortar will completely fill the trowel blade.

5.1.3   Picking Mortar From A Pan

Try this method when the mortar is in a pan:

Step 1   Rotate the edge of the trowel down and use it to cut a slice from the near edge of the pan to the far edge.

Step 2   Cut back toward the near edge, angling outward with the edge of the trowel. Do not turn the trowel over and do not lift the trowel out of the pan while making the return cut. This marks a long, thin wedge of mortar shaped like a steeple. If the mortar looks like a piece of pie, it cannot be picked up properly. *Figure 27* shows the shape of the cut mortar.

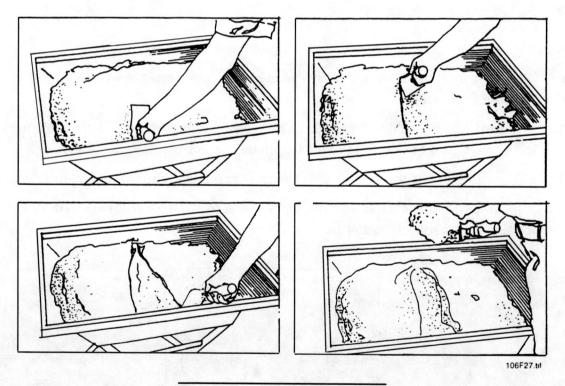

106F27.tif

Figure 27. Cutting Mortar In A Pan

Step 3   Without removing the trowel from the mortar, slide the trowel under the mortar so the blade becomes parallel to the floor. Or, in other words, scoop the trowel under the mortar at the edge of the pan.

Step 4   Firmly push the trowel, with the blade parallel to the floor, toward the middle of the pan. The mortar will pile up on the blade.

Step 5   Lift the trowel from the mortar at the end of the stroke. The trowel should be fully loaded with a tapered section of mortar.

Step 6   To prevent the mortar from falling off the trowel, snap your wrist slightly to set the mortar on the trowel as you lift. This needs to be a very small snap, or you will snap the mortar off the trowel. If that happens, start over.

5.2.0   SPREADING, CUTTING, AND FURROWING

The next sections describe spreading the mortar, shaping its edges, and **furrowing** it. You can practice spreading, cutting, and furrowing the mortar along a 2 × 4 board spread between two trestles, cement blocks, or other props. Practice until you feel comfortable with these movements.

5.2.1   Spreading

Spreading the mortar is applying it in a desired location at a uniform thickness. Mortar is spread for bed joints. The process of spreading the mortar for the bed joint is also called **stringing** the mortar. The spreading motion has two components to it, and they occur at the same time.

The first component of the spreading motion is a horizontal sweep from the starting point or the point where the last **spread** of mortar ended, back toward you. The mortar deposited is called a spread. Try to make the spread about two bricks long to begin with. If you are working with block, try to string the spread about one block long at first. After practice, you should be able to string the spread three to four bricks long, or two blocks long.

The second component of the motion is a vertical rotation. The trowel starts with its blade horizontal. As you move the trowel back toward you, you are also rotating it. As you rotate it, your thumb moves downward, and the back of the blade moves away from you. *Figure 28* shows this rotation of the trowel. Throwing the trowel empties, or throws, the spread. As it spreads, the trowel rotates through 180 degrees as your arm moves back toward you.

As the blade tilts, the mortar is forced to leave the trowel. As the blade tilts, the trowel is traveling horizontally, its heel moving toward you. The two motions together form almost a spiral, depositing the mortar along the path of the spiral.

Practice spreading until you can deposit a trail rather than a mound of mortar. Keep the trowel in the center of the wall for the length of the spread, so mortar will not get thrown on the face of the masonry. Start with a goal of 16 inches and work up to a spread of 24 to 32 inches.

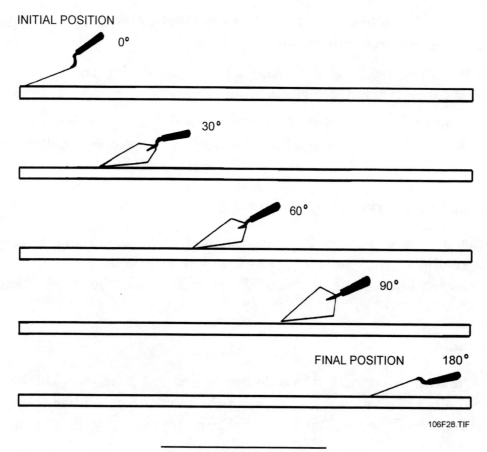

INITIAL POSITION
0°

30°

60°

90°

FINAL POSITION   180°

106F28.TIF

Figure 28. Trowel Rotation

The joint spread should be about ¾ of an inch tall for brick and 1½ inches tall for block. Full joint spreads are used for all brick but not for all block. Usually block is mortared on its face shells and not its webs. However, block needs a full bed joint when it is:

- The first or starting course on a foundation, footing, or other structure
- Part of masonry columns, piers, or pilasters designed to carry heavy loads
- In a reinforced masonry structure, where all cores are to be grouted

Check the specifications to be sure. After the first course, the remaining block is mortared on shells, or shells and webs, according to specifications.

You will need to know how to spread a full bed joint, cut it, and furrow it, whether you work with block or brick.

5.2.2   Cutting Or Edging

After each spread, use the edge of the trowel to cut off excess mortar. To cut, hold the edge of the trowel at about a 60 degree angle, perpendicular to the edge of the mortar. Use the edge of the trowel to shave off the edge of the mortar. *Figure 29* shows the correct angle for shaving the edge of the spread.

Figure 29. Cutting The Edge

Keep the edge of the trowel at a flat angle as shown. This will allow you to catch the mortar as you shave the edge. At this stage in your practice, learn how to catch the mortar as you cut it. The excess mortar can be returned to the mortar pan or used to fill any spaces in the bed joint.

Catching the mortar as you shave it means you do not have to go back and pick it up afterwards. On the job, acceptable work does not include mortar stuck to the face of the masonry unit or lying in piles at the foot of a wall. Mortar is hard to remove when it dries, easy to clean when it is fresh. Learn to clean mortar as you lay it.

5.2.3 Furrowing

Furrowing is the act of shaping the bed joint before laying a masonry unit on it. A furrow is a shallow triangular depression, like a trough, extending the length of the bed joint. The furrow gives the mortar room to move slightly, just enough to let the mason adjust the masonry unit to its proper position. If the furrow is too shallow, the masonry unit will not move easily. If the furrow is too deep, it may expose the unit below and eventually cause a leak.

To make the furrow, hold the trowel blade at a 35 degree angle to the length of the spread, with the point into the spread. The point of the trowel should not go below the depth you want the finished furrow to stand. Tap the trowel point into the mortar at that angle and repeat the taps along the length of the spread. *Figure 30* shows the furrow; notice the overlaid spacing for the trowel taps.

106F30.tif

Figure 30. Furrow And Spacing

After furrowing the length of the spread, cut back the excess mortar. Use the edge of the trowel blade to shave off excess mortar hanging over the face of the wall. As you shave it, catch the excess mortar on the trowel blade. Use the excess mortar to **butter** the **head joint** on the next masonry unit to be laid.

5.3.0 BUTTERING JOINTS

Buttering the head joint is applying mortar to a header surface of a masonry unit. Buttering occurs after the bed joint is spread and the first masonry unit is laid in the bed. Buttering techniques are different for brick and block.

5.3.1 Buttering Brick

Buttering brick is a two-handed job. Begin by spreading the mortar on the bed joint. Keeping the trowel in your hand, pick up the first brick with your other hand. Press this brick into position in the mortar. Cut off the excess mortar on the outside face with the edge of the trowel.

Keeping the trowel in your hand, pick up a second brick in your brick hand. As you hold it, apply mortar to the header end of the brick. *Figure 31(C)* shows a properly buttered head joint.

Notice that the buttered mortar covers all the header surface but does not extend past the edges of the brick. Hold the trowel at an angle to the header surface to keep the mortar off the sides of the brick.

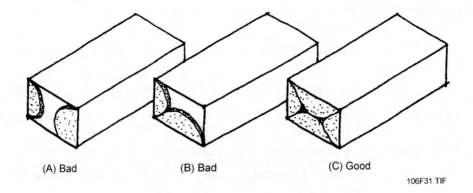

(A) Bad   (B) Bad   (C) Good

106F31.TIF

Figure 31. Buttered Head Joint

When the brick is buttered, use your brick hand to press it into position next to the first brick. You can press your palm across the ends of both bricks, as shown in *Figure 32*. This will let you feel when the two bricks are even. Give it an eyeball check as well. After placing the brick, cut off the excess mortar with the edge of your trowel.

106F32.tif

Figure 32. Placing The Second Brick

Unlike blocks, bricks are easily held in one hand. Take advantage of this to use both hands for laying bricks. Try to develop a rhythmic set of movements. This will make the work faster and easier on you. Remember to use your shoulders and arms, not just your wrists.

After you have laid six bricks, check them for placement. Use your mason's level to check both plumb and level. If a brick is out of line, tap it gently with the handle of your trowel. Do not tap the level. Do not use the point or blade of your trowel or it will lose its edge.

5.3.2 Block Head Joints

Buttering block is more complicated than buttering brick. Block is larger and heavier than brick and not an easy one-handed lift. Use two hands when lifting block to avoid strain. Block is more demanding than brick in that both blocks must be buttered to get a good head joint. Block is also more demanding than brick in that it calls for three different types of bed joints.

Again, start by spreading and furrowing a bed joint. Position the first block in the mortar. Then stand two or three blocks on end next to their bed. Since block is wider at the top than at the bottom, stand the block so that the top sides will be on top when the block is placed. This makes it quick to butter several blocks at once.

You do not need to fill the trowel with mortar because one block does not take as much mortar as one brick. Butter the ear ends of the standing blocks. Wrap the mortar around the inside of each ear to help hold it in place. Then butter the ear end of the laid block. Then lift the standing block by grasping the webs, or ends, with both hands. As shown in *Figure 33*, keep the trowel in your hand to save time. Do not jerk the block, or the mortar will fall off.

106F33.tif

Figure 33. Placing A Block

Place the block against the buttered, laid block. Tilt the block slightly toward you as you lay it into place, so that you can see the alignment of the cores and edges. Use your eyes to check that the edge of the block aligns with the block directly below.

To seat the block, gently press down and forward, so that the mortar squeezes out at the joints. Do not drop the block, but ease it into place. Continue laying the pre-buttered blocks, being sure to butter the ear end of the laid block each time.

An alternative method of buttering block is to butter one end, lift it by the webs with one hand, and butter the other end. This method is not recommended for beginners.

After you place the block, cut off the excess mortar with the edge of your trowel. Check for level and plumb with your mason's level. Use the handle of your trowel to gently tap the block into place. After the block is placed, the mortar joint spacing should be the standard ⅜ of an inch for both the bed and head joints.

Do not move the block after it is pushed against its neighbor. If you must move the block, take it off and remortar the bed joint and the head joint. Unlike brick with its solid and complete mortaring, block mortaring is fragile. Because its webs are so small in area compared to its size, block mortar joints are easily disturbed by movement. Do not take the chance of a weakened mortar joint developing a leak in the wall.

5.3.3   Block Bed Joints

While bricks have one type of bed joint, the full furrowed joint, blocks can use one of three types of bed joint depending on their purpose. Check the specifications before laying a block wall to confirm which type of bed joint to use.

- If the block is laid as the first course on a footing, it takes a full furrowed bed joint, as does brick.

- If the block is not to be in a reinforced wall, the bed joint has mortar on the face shells only. *Figure 34* shows this type of mortaring.

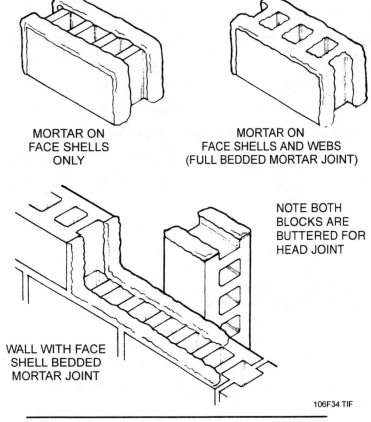

MORTAR ON
FACE SHELLS
ONLY

MORTAR ON
FACE SHELLS AND WEBS
(FULL BEDDED MORTAR JOINT)

NOTE BOTH
BLOCKS ARE
BUTTERED FOR
HEAD JOINT

WALL WITH FACE
SHELL BEDDED
MORTAR JOINT

106F34.TIF

Figure 34. Face Shell And Full Mortar Bed Joints

- If the block is part of a reinforced wall that will have grout in the cores, the block needs a full block bed joint. This has mortar on the face shells and on the webs, as shown in *Figure 34*. Mortaring the webs will keep the grout from oozing out of the cores.

Sometimes, the specifications will call for an unreinforced wall to be laid with a full block bed joint. Mortaring the webs as well as the shells increases the loadbearing strength of the wall. The architect or engineer may have calculated that a full block bed joint will do the job instead of reinforcement. If you use only a shell bed joint, the wall will not have the calculated strength. This is another reason why it is important to read the specifications.

5.4.0 GENERAL RULES

The way you work the mortar determines the quality of the joints between the masonry units. The mortar—and the joints—form a vital part of the structural strength and water resistance of the wall. Learning these general rules and applying them as you spread mortar will help you build good walls:

- Use mortar with the consistency of mud so it will cling to the masonry unit, creating a good bond.
- Butter the head joints thoroughly for brick and block; butter both sides of the head joints for block.
- When laying a unit on the bed joint, press down slightly and sideways, so the unit goes against the one next to it.
- If mortar falls off a moving unit, replace the mortar before placing the unit.
- Put down more mortar than the size of the final joint; remember that placing the unit will compress the mortar.
- Do not string a spread more than 6 bricks or 3 blocks long; longer spreads will get too stiff to bond properly as water evaporates from them.
- Do not move a unit once it is placed, leveled, plumbed, and aligned.
- If a unit must be moved after it is placed, remove all the mortar on it and rebutter it.
- After placing the unit, cut away excess mortar with your trowel and put it back in the pan, or use it to butter the next joint.
- Throw away mortar after 2 to 2½ hours, as it is beginning to set and will not give a good bond.

6.0.0 BONDING MASONRY UNITS

Masons deal with four types of bond.

- A *simple mechanical bond* is made by the joining of mortar and a masonry unit. The strength of this bond depends on the mortar. This is also called a mortar bond.
- A *pattern bond* is a pattern formed by masonry units and mortar joints on the face of a surface. Unless it is the result of a structural bond, a pattern bond is purely decorative.

- A *structural bond* is made by interlocking or tying masonry units together so they act as a single structural unit.
- A *structural pattern bond* is the result of a structural bond that forms a pattern as well as a bond. Most traditional pattern bonds are structural pattern bonds.

The next sections discuss these different types of bonds. Note that the distinction between a structural bond and a pattern bond is hard to make. The act of overlapping or interlocking masonry to create a structural bond also creates a pattern. Defining a particular pattern as a structural bond or a structural pattern bond depends on local custom.

6.1.0 MECHANICAL OR MORTAR BOND

On the basic level, a mechanical bond is formed between the masonry unit and the mortar. This bond ties the masonry in a wythe into a single unit.

For most masonry construction, the most important property of mortar is bond strength. Mortar bond strength depends on the properties of the mortar and of the bonding surface.

- The mortar must have the right proportions of ingredients for its use. It must stay wet enough to lay and level the masonry.
- The masonry surface should be irregular to provide mechanical bonding. It should be absorptive enough to draw the mortar into its irregularities.
- The masonry surface should not be so dry that it dries out the mortar. Slow, moist curing improves mortar bond and compressive strength.

The second most important property of mortar is bond integrity. Bond integrity depends on the mason who:

- Keeps tools and masonry units clean
- Butters every joint fully without air bubbles
- Does not move the masonry unit after it is leveled
- Levels units shortly after they are laid
- Uses mortar wet enough to dampen the masonry unit
- Keeps the mortar tempered
- Mixes fresh mortar after 2 hours

The work of the mason defines the bond between masonry units.

6.2.0 PATTERN BOND

Pattern bonds add design but not strength to masonry walls. The stack bond, *Figure 35*, is only a pattern bond. It provides no structural bond as there is no overlapping of units. This pattern is more commonly used with block than brick.

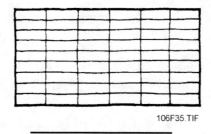

106F35.TIF

Figure 35. Stack Bond

If this pattern is used in a loadbearing wall, the wythe must be bonded to its backing with rigid steel ties. In loadbearing construction, this patterned wall should be reinforced with steel joint reinforcement ties.

Pattern and structural pattern bonding calls for placing brick in different positions in the wythes. *Figure 36* shows different ways of placing brick in order to make different kinds of patterns.

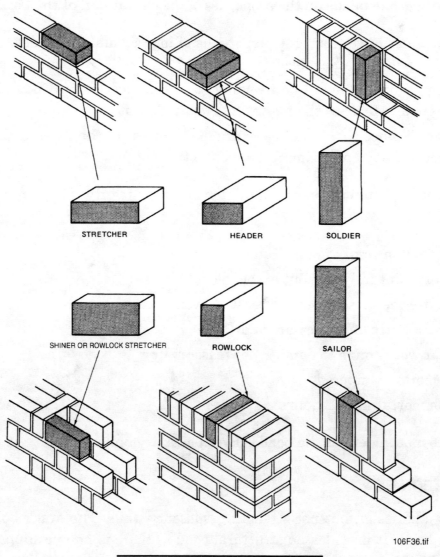

106F36.tif

Figure 36. Brick Positions In Wall

The stretcher is the everyday workhorse. Headers are used primarily for tying wythes together, capping walls, flat windowsills, and pattern bonds. Soldiers are used over doors, windows or other openings, and in pattern bonds. Shiners, or rowlock stretchers, are used in pattern bonds, in brick walks, and for leveling when a 4-inch lift is needed. Rowlocks are found in capping walls, windowsills, ornamental cornices, and pattern bonds. Sailors are rarely seen, except in pattern bonds and brick walks.

6.3.0   STRUCTURAL BOND AND STRUCTURAL PATTERN BOND

Wythes can be structurally bonded by using metal ties, joint reinforcements, anchors, grout, and steel rods. These are engineering methods used to increase strength and loadbearing by firmly tying masonry units and wythes together.

Another, older way to structurally bond a wythe is to lap masonry units. Lapping one unit halfway over the one under it provides the best distribution of weight and stress.

In a single-wythe wall, a structural bond is made by staggering the placement of the bricks. This results in the brick in one course overlapping the brick underneath. The structural pattern bond resulting from this simple overlap is the running bond, as shown in *Figure 37*. Common overlaps are the ½ lap and the ⅓ lap. Changing the proportion of the overlap changes the look of the pattern, as you can see in *Figure 37*.

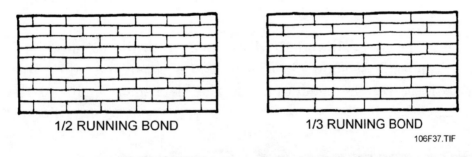

1/2 RUNNING BOND   1/3 RUNNING BOND

106F37.TIF

Figure 37. Common Running Bonds, ½ And ⅓ Overlaps

In two-wythe walls, a structural bond is made between the wythes. This can be made by rigid steel ties that equalize loadbearing. It can also be made by overlapping a brick from the face wythe to the backup wythe. The overlap brick is turned into the header or the rowlock position. This results in a complex structural bond which is also a structural pattern bond, with different sizes of brick facing out. The results are the traditional English bond and Flemish bond shown in *Figure 38*.

The Flemish bond consists of alternating headers and stretchers in every course. The English bond consists of alternating courses of headers and stretchers. If the headers are not needed for structural bonding, cut bricks are used. Brick can be laid to show different faces and cut in different ways.

Note that the corners in *Figure 38* use different patterns. The English corner uses brick in the rowlock position and half bricks. The Dutch corner uses half bricks. English and Dutch corners are the most commonly used corner patterns. Sometimes these and other corner patterns are used around door and window openings as well.

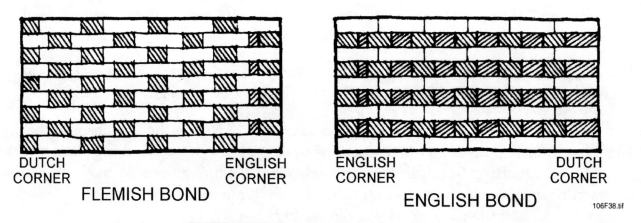

DUTCH CORNER   ENGLISH CORNER

FLEMISH BOND

ENGLISH CORNER   DUTCH CORNER

ENGLISH BOND

106F38.tif

Figure 38. Flemish And English Bonds

The combination of the Flemish and English bonds with the running bond results in the common or American bond. As shown in *Figure 39*, the common bond is a running bond with headers every sixth course. The headers are in the Flemish or English pattern, according to the specifications.

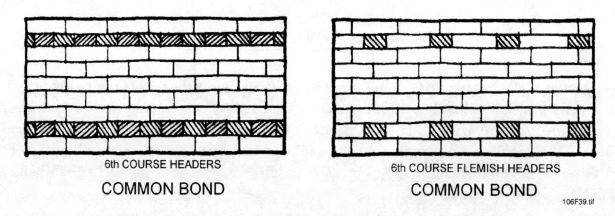

6th COURSE HEADERS

COMMON BOND

6th COURSE FLEMISH HEADERS

COMMON BOND

106F39.tif

Figure 39. Common Bond, With English And Flemish Headers

The English cross, or Dutch bond, uses a structural pattern bond that repeats every four courses. The pattern courses are all stretcher, all header, and a course of three headers and one stretcher. The last pattern course is all header again. *Figure 40* shows the Dutch bond.

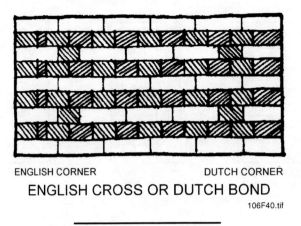

ENGLISH CORNER   DUTCH CORNER
ENGLISH CROSS OR DUTCH BOND
106F40.tif

Figure 40. Dutch Bond

The Dutch bond may seem complicated until you look at a traditional garden wall bond. *Figure 41* shows two variations on the garden wall structural pattern bond. The double stretcher garden wall pattern repeats every five courses. The dovetail garden wall pattern repeats every 14 courses. More variations can be seen and imagined. The only limit on patterning is the skill and ingenuity of the mason.

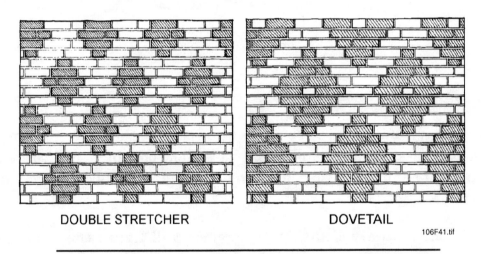

DOUBLE STRETCHER   DOVETAIL
106F41.tif

Figure 41. Double Stretcher And Dovetail Garden Wall Bonds

6.4.0 BLOCK BOND PATTERNS

Block has its own set of commonly used bond patterns. *Figure 42* shows common block bonds, some of which, such as the herringbone, are also seen in brickwork. As with brick, the stack bonds do not provide any structural strength. With block, however, it is simple to reinforce stack bond with grout and steel or grout in the cores.

The other block bond patterns add structural strength. To get a solid face, block can only be laid in the stretcher mode. Pattern variations, such as the coursed ashlar, can be made by using different sizes of block. Modern block walls can also add visual interest through texture and surface designs. Many designers rely on textures and surface designs, alone or in combination with bond patterns, to enhance block walls.

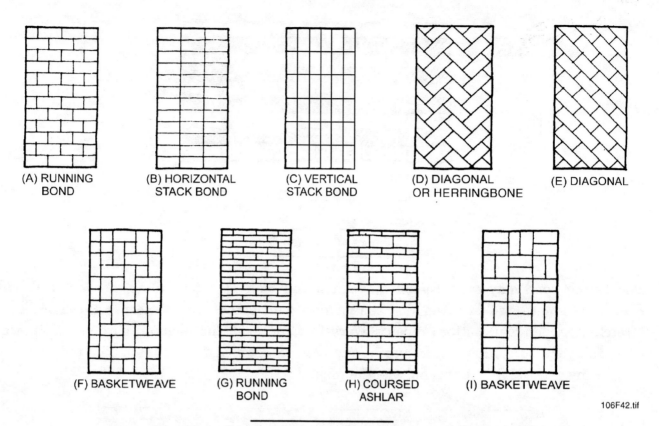

(A) RUNNING BOND

(B) HORIZONTAL STACK BOND

(C) VERTICAL STACK BOND

(D) DIAGONAL OR HERRINGBONE

(E) DIAGONAL

(F) BASKETWEAVE

(G) RUNNING BOND

(H) COURSED ASHLAR

(I) BASKETWEAVE

106F42.tif

Figure 42. Block Bonds

7.0.0   CUTTING MASONRY UNITS

Many times, masonry units need to be cut to fit a specific space. Even when building on a modular grid, structural bond patterns, door and window openings, and corners usually call for some cut masonry units. English and Dutch corners specifically call for cut masonry units as part of the patterning.

On a large job, the masonry contractor or foreman will figure the pattern layouts and calculate the number of masonry units to be cut. Someone will be assigned to cut the units with a masonry saw or a splitter before they are needed. Sometimes, masons need to cut a few more units or cut to a slightly different size. This is when you need to know how to cut masonry with hand tools.

7.1.0   COMMON CUT SHAPES

Both brick and block are cut in standard cuts. The next sections describe these cuts for brick and block.

7.1.1   Brick Cuts

Brick is easily cut by hand tool, masonry saw, or splitter. Sometimes cut bricks are needed for finishing corner patterns or for pattern bonds. *Figure 43* shows the common cut brick shapes and names. The king and queen closures are used for cornering.

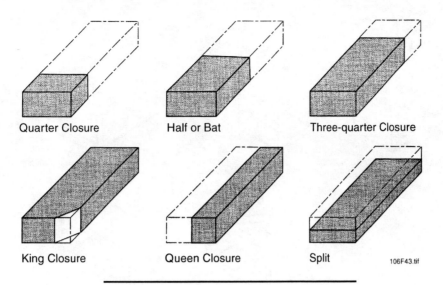

Figure 43. Cut Brick Shapes And Names

Cut brick is sometimes needed for building a pattern bond in a veneer or cavity wall. *Figure 44* shows a cavity wall with cut rowlock and header bricks. Because of the way they are turned, the soldier bricks do not need to be cut.

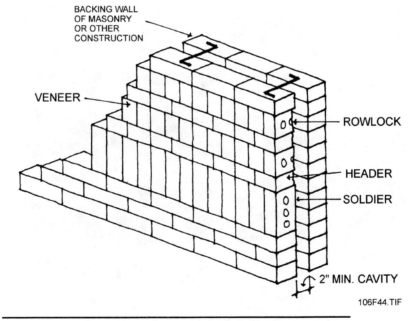

Figure 44. Cut Rowlocks And Headers In A Pattern Bond

7.1.2 Block Cuts

Block is usually cut in several standard ways. It can be cut across the stretcher face, both horizontally and vertically, as shown in *Figure 45*. These cuts are easily made by hand. Blocks cut across the face horizontally are called splits or rips. If the block is cut exactly in half, it is called a half-high rip. Rip blocks are often used under windows. They act as a filler to reach a height of 8 inches so normal coursing can continue.

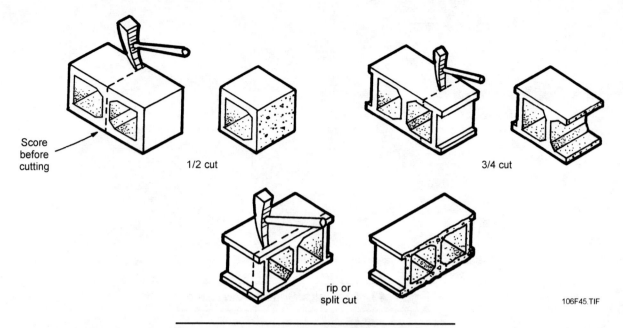

Figure 45. Horizontal And Vertical Face Cuts

Blocks also get their webs cut out. Taking one end off a block makes an opening easily slipped over a pipe. Fitting the block to its location may take more cuts. Standard internal cuts for block are shown in *Figure 46*. These cuts are usually made by masonry saw. The cuts have their own names to save time and confusion.

Figure 46. End And Web Cut Blocks

The three types of cuts that are shown in *Figure 46* are:

*   The quarterback cut has the end of one cell cut out, leaving two cells.
*   The halfback cut has the end of one cell and the web of the next cell cut. This leaves one cell.
*   The fullback cut has one end and both internal webs cut. This leaves only one end to hold the block together.

If a block has only two cells, the cuts are halfback and fullback; there is no quarterback cut for a two-cell block.

The bond beam block has the ends and inside webs of the block cut down about three fourths of the way. This cut, in *Figure 47*, can be used for a lintel over an opening. The cuts give room for the reinforcement on top of the opening.

106F47.TIF

Figure 47. Bond Beam Cut

7.2.0   CUTTING WITH HAND TOOLS

Brick and block can be cut with chisels or a brick hammer. Brick can be cut with the edge of a trowel. The procedures are detailed in the following sections.

CAUTION: Remember to wear a hard hat and eye protection when cutting with hand tools. Never cut masonry over the mortar pan or near other workers. Chips may fly off, hitting other workers.

7.2.1   Cutting With Chisels And Hammers

Using the chisel and hammer can result in a smooth cut for block and brick. This procedure works well for both types of units.

Step 1   Check the tools you will use. Cutting edges should be sharp, and the hammer handle should be firmly attached.

Step 2   Put on your hard hat and safety goggles or other eye protection.

Step 3   Put the brick or block on a bag of sand, a board, or the earth to make a safe cutting surface. Make sure it is resting flat and plumb on a surface with some give to it.

Step 4   Use a steel square and a pencil to mark the cut all the way around the masonry unit.

Step 5   Hold the blocking chisel (for blocks) or the brick set (for bricks) vertically on the marked line. The flat side of the chisel should face the finished cut, or the part you want to keep.

Step 6   Give the chisel end several light taps with the striking end of the hammer to score the masonry unit. Move the hammer and chisel all around the unit, scoring all along the cut mark. Be sure to keep your fingers above the cutting edge of the chisel. *Figure 48* shows this step.

Step 7   Turn the unit so the finished cut is toward you and the waste part is away from you.

Step 8   Place the chisel on the scored line, with the flat side facing the finished cut. Deliver a hard blow to the chisel head with the hammer. Sometimes two blows are needed.

106F48.tif

Figure 48. Scoring Brick With The Brick Set

Cutting in this way gives an accurate and clean cut. An accurate cut can also be made if a hammer is used for the final step instead of the chisel. If you are cutting with the mason's hammer, follow Steps 1 through 7 above, then continue with these steps:

Step 1   Place the scored block on top of another block so that the waste part **hangs free**. You can hold the wanted part secure with your foot.

Step 2   Strike the waste end of the block with the striking end of the hammer. **This knocks** off the waste end, leaving a clean finished cut.

7.2.2 Cutting With Masonry Hammers

Cutting with the chisel end of the masonry hammer gives a rougher cut. The steps for cutting brick in this way are as follows:

Step 1   Check the tool you will use. Cutting edges should be sharp, and the hammer handle should be firmly attached.

Step 2   Put on your hard hat and safety goggles or other eye protection.

Step 3   Use a steel square and a pencil to mark the cut all the way around the masonry unit.

Step 4   Hold the brick in one hand and your hammer in the other. Hold the part of the brick you want to keep with the waste part down.

Step 5   Strike the brick lightly with the chisel end of the hammer to score it along the marks on all sides. As you turn the brick, be sure to keep your fingers and thumb off the side of the brick being scored. *Figure 49* shows this step.

Step 6   Strike the face of the brick sharply with the chisel end of the hammer. Let the waste part fall to the ground.

Step 7   If necessary, use either end of the hammer to dress out any small, rough edges left by the cut.

106F49.tif

Figure 49. Scoring Brick With The Hammer

The same procedure can be used for block, except that block is not usually held in your hand. Set the block on sand, earth, or a board for a safe cutting surface. Follow Steps 1, 2, 3, and 5 above. Then tilt the block face away and prop it with another block. Hold it with your foot and apply a sharp blow with the hammer. Blocks may need to be struck on both faces. Finish by dressing out any rough edges.

7.2.3 Cutting With Trowels

Cutting with a trowel is not recommended for block or very hard brick. Cutting with a trowel in cold weather can break the blade.

Cutting with a trowel is a last resort when your hammer is not available. After you have mastered cutting with the hammer and brick set this will be easier to learn.

Mark the brick for cutting. Hold the brick in one hand by the part you want to keep. Keep your fingers well under the brick, as in *Figure 50*, to avoid cutting them. Strike the brick using the upper edge of the trowel, close to the heel. Strike hard with a quick sharp blow and a sharp snap of your wrist. If the brick does not break, use a brick hammer and follow the steps above.

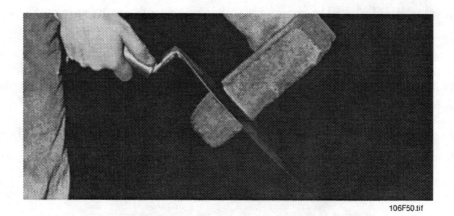

106F50.tif

Figure 50. Cutting With The Trowel

7.3.0 CUTTING WITH SAWS AND SPLITTERS

Cutting masonry units with power saws or splitters takes two kinds of awareness. You must be aware of how to operate the machinery, and also be aware of the masonry units and cuts.

7.3.1 Saws And Splitters

Masonry saws are freestanding or portable models. They use either diamond or carborundum blades. Diamond blades are irrigated to prevent fire. The water wets the masonry unit, which must dry out before it can be laid. Carborundum blades are not irrigated, but they make clouds of dust. The dust must be blown or vented away from the saw and nearby workers. Smaller hand-held saws use dry blades and also make clouds of dust which must be blown or vented away.

Splitters do not use water or generate dust. They do, however, exert tremendous force through gearing and hydraulic power.

As with any potentially dangerous equipment, follow these general safety rules:

- Do not operate any saw or splitter until you have had specific instructions in handling that equipment.
- Check the condition of the equipment before using it.
- Follow all safety rules for using power equipment.
- Follow all safety rules for using dangerous equipment.
- Wear a hard hat, respiratory mask, goggles, gloves, and other Appropriate Personal Protective Equipment as needed.
- Never force the equipment.
- For bedded saws, use conveyor carts, pushers, or blocks to move the unit under the blade.
- For hand-held saws, secure and brace the unit before cutting it.
- Do not operate equipment when you are feeling ill.

Review the safety rules as well as the operating instructions before operating any equipment.

7.3.2 Units And Cuts

After you have checked out the equipment, the safety procedures, and the operating procedures, check out the masonry units.

- Know what the finished item should look like.
- Mark all cutting lines before the blade starts running.
- Mark cutting lines in grease pencil for wet-cut saws.
- Do not cut a cracked masonry unit.

If you are not clear about the cuts to be made, ask your supervisor for more direction.

8.0.0 LAYING MASONRY UNITS

Laying masonry units is a multi-stage process. As discussed in previous sections, the first step in any masonry job is reading the specifications. This is followed by planning the layout of the job. The next tasks are to locate and lay out the wall, then do the dry bonding. Dry bonding assures that the layout will be correct and that the minimum number of cut blocks will be needed. Then, calculate the number of units to cut, and cut them. For the purpose of this module, assume all work is done in running bond on the modular grid system. Now you are ready to mix the mortar and start the actual laying.

The next tasks are to spread mortar, lay masonry units in place, and check their positioning.

An earlier section gave detailed procedures for spreading and furrowing bed joints and buttering head joints. The following sections give procedures for positioning individual masonry units, laying to the line, and building corners and **leads**.

8.1.0 LAYING BRICK IN PLACE

Laying brick will be less of a strain if you use as few motions as possible. One way to make things easier is to have your materials close by. If you are working on a veneer wall, use both hands to pick up bricks and stack them on the completed back section of the wall. Place them along the length you will be laying before you start spreading mortar. This will eliminate the need to bend to pick bricks off the ground as you go.

Another efficient practice is to use both hands. Keep your trowel in one hand and use the other for picking, holding, and placing bricks. This will make the work easier and faster.

Use your fingers efficiently as well. When you pick up a brick, hold it plumb. Pick it up so that your thumb is on the face of the brick. Let your fingers and thumb curl down over the top edges of the brick, slightly away from the face. Your fingers in this position will not interfere with the line as you place the brick on the wall. *Figure 51* shows the proper position of fingers and thumb for holding and laying the brick in place.

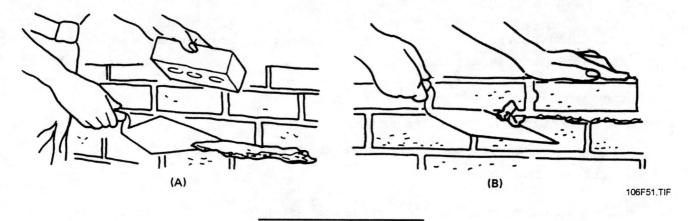

(A)   (B)

106F51.TIF

Figure 51. Finger Positions

Keep your mason's level close by. After laying every six bricks, check them for position. This means checking for height, level, plumb, and straightness as you go. If you cannot adjust a unit to meet the four measures of height, level, plumb, and straightness, you must take the brick out and start over.

8.1.1 Placing Brick

The most important placing rule is to place gently. Do not drop the brick or block onto the bed joint. Lower it down gently, as shown in *Figure 52*. Press it forward at the same time, so that it will butt against the unit next to it. Mortar should ooze out slightly on both head and bed joints to show that there has been full contact.

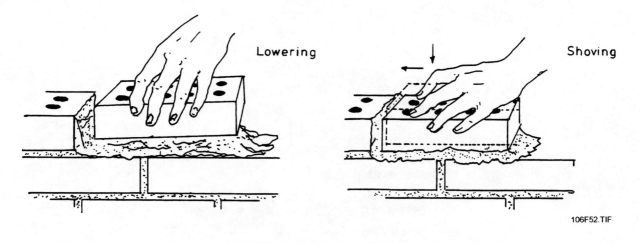

Figure 52. Placing The Brick

Align the latest brick with the brick next to it as you place it. Line it up with the mason's line if you are using one. By standing slightly to one side, you will be able to sight down the wall. This will help maintain plumb head joints by sighting the brick below the newly laid unit.

You may need to slightly adjust the unit in its bed. First try pressing downward on the brick with the heel of your hand, as in *Figure 51(B)*. Keep part of the heel of your hand on the brick next to the one you are adjusting. If this is not sufficient, you may need to tap the brick with the heel of your trowel. After adjusting the brick, cut off the extruded mortar with your trowel and lay the next unit. Cutting mortar as you go will help you to keep the masonry clean.

8.1.2   Check Height

The first check is always course height. If this is off, there is no use checking anything else. Use your modular or standard course spacing rule to check the height of the brick.

- After the bricks are laid on the wall, unfold the rule and place it on the base or footing used for the mortar and brick.

- Hold the rule vertically, as in *Figure 53*. Check that the end of the rule is flat on the base, so the reading is accurate. If you are using standard brick, the first course should be even with number 6 on the modular rule. If you are using a different size of brick, check the appropriate scale on the modular or course rules.

- If the height of the course does not hit the right place on the rule, take the bricks out and clear off the bed joint. Lay the bed joint again and replace the buttered bricks. Recheck the height of the course. Then replumb and relevel.

The height, or vertical course spacing, depends on the thickness of the mortar joints. Practicing laying full bed and head joints is the fastest way to learn to make standard size joints.

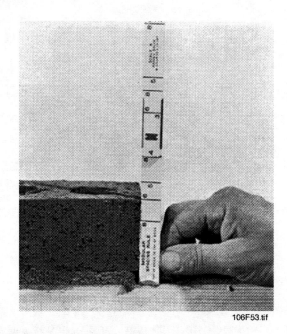

Figure 53. Modular Rule

If more than one course is laid, always set the modular rule on the top of the first course to measure. The base may have been irregular, and a large joint may have been used to level the first course.

8.1.4 Check Level

After checking the height for your string of six bricks, check with your mason's level for levelness.

- Remove any excess mortar on top of the bricks.
- Place your mason's level lengthwise on the center width of the six bricks to be checked.
- Use your trowel handle to gently tap down any bricks that are high with relation to the mason's level. Do not tap them so hard they sink low.
- If bricks are low, pick them up. Clean and mortar the bed and the head joint again and reposition the brick. Reposition the mason's level again and get it level.

8.1.4 Check Plumb

The next step after leveling is to check for plumb, or vertical straightness.

- Hold the level in a vertical position against the end of the last brick laid.
- Tap the brick with the trowel handle to adjust the brick face either in or out, **as in** *Figure 54.*
- Move the level to the end of the first brick laid and repeat the process.

Figure 54. Plumbing The End Bricks

Figure 55 gives profiles and names for bricks that are plumb and out of plumb. The large black dot represents the mason's line. By looking and touching, you can train your hand and eye to know bricks that are plumb and bricks that are not plumb.

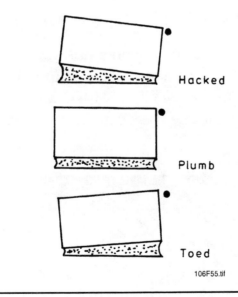

Figure 55. Plumb And Not Plumb Bricks

8.1.5   Check Straightness

After the first and last bricks in the string have been plumbed, check the rest for straightness.

*   Hold the mason's level in a horizontal position against the top of the face of the six bricks, as shown in *Figure 56*.

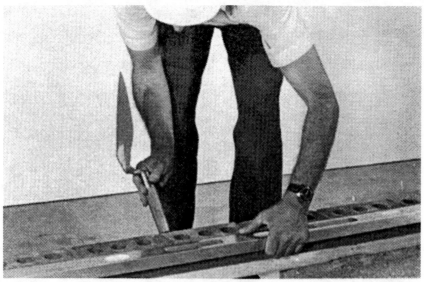

Figure 56. Checking For Straightness

- Tap the bricks either forward or back until they are all aligned against the mason's level. Be careful not to move the end plumb points while you are aligning the middle four bricks.

By sighting down from above you can train your eye to know bricks that are straight and bricks that are not straight.

8.1.6 Laying The Closure Unit

The last unit in a course is called the **closure unit**. Masons lay corners of a wall first, then work from each corner toward the middle. The last unit, or closure unit, must fit in the gap between the masonry units that have already been laid (*Figure 57*). The closure unit should fall toward the middle of a wall. The space left for it should be large enough for the unit and its two head joints.

Figure 57. Closure Unit

The process for laying the closure unit is the same for block and brick. In both cases:

- Butter the closure unit on both head joints and on the bottom bed joint.
- Butter the adjacent units on their open head joints.
- Gently ease the unit into the space, as shown in *Figure 57*.
- If any mortar falls out of a closure unit joint, remove the unit and reset it in fresh mortar.

If the head joints have been properly spaced, the closure unit will slide in with the specified joint spacing. Otherwise, the closure unit will have head joints that are too large or too small. If this is the case, remove the last three or four units that were laid on either side of the closure unit. Remortar them and relay them to correct for the closure head joint size. The objective is to avoid a sudden jump in the size of a head joint. A big change in joint size will catch the eye and can also skew the pattern bond. If you must move bricks, be sure to check them again for height, level, plumb, and straightness.

8.2.0 PLACING BLOCK

Placing block is similar to placing brick, except that it is not a one-handed job. Butter the units on both sides of head joints. Use both hands to place the block. Do not drop it, but move it slowly down and forward so it butts against the adjacent unit. Slightly delaying release allows the block to absorb moisture from the mortar, which makes a good mechanical bond. If the mortar does not ooze from the joints, you are not using enough mortar. There will be voids in the joints, and the wall will eventually leak.

Tilt the block toward you as you position it. Look down over the edge and into the cores to check alignment with the block underneath. If a block is set unevenly, check to see if a pebble or other material is wedged between the mortar and the block. If so, take the block up and clean it. Rebutter fresh mortar for both bed and head joints, and reset it.

If the block requires adjustment, lightly tap it into place. You may want to use your mason's hammer to tap the block as it may require stronger taps.

Check each block for position as described above for brick. If the block cannot be adjusted, take it up, remortar it, and reset it.

8.3.0 LAYING TO THE LINE

To keep masonry courses level over a long wall, masons lay the units to a line. Working to a line allows several masons to work on the same wall without the wall moving in several directions. The line is set up between corner poles or corner lead units. The poles or leads must be carefully checked for location, plumb, level, and height. The line is placed on the outside of the course, so that will be the most precisely-laid side. The mason usually works on the same side as the line, but can also work from the other side depending on job conditions and experience.

The masons place each masonry unit with its outside top edge level with and $\frac{1}{16}$ of an inch away from the line. The distance is the same for all types of masonry units. Your eye will get trained to measure that distance automatically after some practice laying to the line.

8.3.1 Setting Up The Line

Mason's line needs to be tied to something that will not move. It needs to be tied taut at a height that can be measured precisely. Mason's line is attached to corner poles or corner leads by means of line stretchers, line blocks, or line pins.

• *Using Corner Poles* – Corner poles allow masons to lay to the line without laying the corners first. Attach the corner pole securely. It must not move as you pull a mason's line from it. For brick veneer walls, the corner pole can be braced against the frame or backing wall. It is important that you check the placement of the corner poles before you string the line. If the pole has course markings, check that they are the correct distance from the footing. If the pole has no markings, transfer markings from your course rule. Make sure you start the measures from the footing.

Attach the line to the left pole to start. If the pole has no clamps or fasteners, attach the line with a hitch or half hitch knot. Stretch it to the right pole and gradually tighten it until it is stretched. Use a hitch or half hitch to secure it, tightening it as you tie. Check that it is at the proper height before you start laying.

After laying each course, move the line up to the next course level. Stretch and measure it again. It is critical to make sure the line is at the proper height for each course. Use your modular rule or course spacing rule to check the line height at each end for every course.

• *Placing Line Blocks And Stretchers* – Mason's line can be set between corner leads or corners laid to mark the ends of the wall. The line can be attached by line blocks, line stretchers, or line pins.

Line blocks have a slot cut in the center to allow the line to pass through. It takes two sets of hands to set up line blocks. The procedure is as follows:

- Pass the line through the slot of the block. Tie a knot, or tie a nail on the end of the line, to keep it from passing through the slot.
- One person holds the line block aligned with the top of the course to be laid. Traditionally, mason's lines start on the left side as you face the wall.
- Place the block so that it hooks over the edge of the masonry unit, and hold it snug.
- The second person walks the line to the right end of the wall.
- The second person then pulls the line as tightly as possible, and wraps the line **three or four turns** around the middle of the block.
- The second person hooks the tensioned line block over the edge of the corner.
- Both parties check that the line is at the correct height.

Line stretchers are put in place following the same steps. The line stretcher slips over the top of the corner blocks, not the edges. Line stretchers are useful when the corner lead is not higher than the course to be laid.

• *Using Line Pins* – Steel line pins or 10-penny common nails also hold mason's line in place. The line pin is less likely to pull out of the wall because of its shape. The peg end of the line, or the starting end, is traditionally started at the left as the mason faces the wall.

Drive the peg line pin securely into the lead joint. Make sure that the top of the pin is level with the top of the course to be laid, as shown in *Figure 58*. Place the pin at a 45-degree downward angle, several units away from the corner. This will prevent the pin from coming loose as the line is pulled.

Tie the line securely to the pin using the notches on the pin. Give the line a few very sharp, strong tugs. This tests whether the pin will come out as the line is tightened and helps to prevent accidents caused by flying line pins.

106F58.TIF

Figure 58. Line Aligned With Top Of Brick

Walk the line to the other lead. Drive the second line pin securely into the lead joint even with the top of the course. Check and measure that the pin is secure and in the correct position before applying the line.

Wrap the line around the pin and start tensioning. Pull the line with your left hand and wrap it around the line pin with your right hand. Use a clove hitch or half hitch knot to secure the taut line to the pin. Be careful not to pull the line so tightly that it breaks.

When you move the line up for another course, immediately fill the pin holes with fresh mortar. If you wait until later to fill the line pin holes you will need to mix another batch of mortar. Taking care of the holes as you move the pins saves many steps at the end of the project.

• *Using Line Trigs* – To keep a long line from sagging, masons set trigs to support the line midstring. The first step in setting a trig is to set the trig support unit in mortar in position on the wall. Be sure that this unit is set with the bond pattern of the wall, close to the middle of the wall. Check that the unit is level and plumb with the face of the wall. Check that the unit is at the proper height with a course rule or course pole. Sight down the wall to be sure that the trig unit is aligned with the wall and is set the proper distance from the line. The trig support unit is a permanent part of the wall, so place it carefully.

After the trig support unit is in the proper place, slip a trig or clip over the taut line. Check that the line is still in position. Lay the trig on the top of the support unit with the line holder on the bottom side. Place another masonry unit on top of the trig to hold it in place. *Figure 59* shows a trig lying on top of a support unit secured by an unmortared masonry unit on top of it.

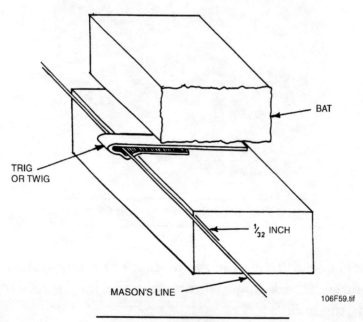

BAT

TRIG
OR TWIG

$\frac{1}{32}$ INCH

MASON'S LINE

106F59.tif

Figure 59. Trig Secured By Bat

Check the line for accuracy once more. The line should just be level with and slightly off the corner of the trig support unit and the standard $\frac{1}{16}$ of an inch away from it.

8.3.2   Laying Brick To The Line

After you string the line and put on a trig, you can begin to lay masonry units to the line. The advantage of laying to the line is that it cuts down on the need for the mason's level.

You must lay to the line without disturbing the line. If the line is tweaked, other masons on the line will have to wait for the line to stop moving. Hitting the line is called **crowding the line**. Even experienced masons will crowd the line occasionally. It is not possible to lay to the line without crowding the line to some degree, but it can be minimized.

To avoid crowding the line, hold your brick from the top, as shown in *Figure 60*. As you release the brick the last quarter inch, the line should pass between your fingertips and the brick. This may take some practice.

106F60.tif

Figure 60. Placing Brick Over The Line

The brick must come to sit $\frac{1}{16}$ of an inch inside the line. The top of the brick must be even with the top of the line, as shown in *Figure 61*. Looking at the brick from above, you should be able to see a sliver of daylight between the line and the brick. Brick set too close is crowding the line. Brick set too far is **slack to the line**. When laid correctly, the bottom edge of the brick should be in line with the top of the course under it, and the top edge of the brick should be $\frac{1}{16}$ of an inch back and even with the top edge of the line.

Adjust the brick to the line by pressing down with the heel of your hand. Check that the brick is not hacked or toed, as shown previously in *Figure 55*. While you are learning to lay to the line, it is a good idea to check your placement with the mason's level. After you have gained some skill in working to the line, you will find you will not need to use the mason's level so often.

With your trowel, cut off the mortar just squeezed out of the joints. Apply the mortar to the head of the brick just laid. By buttering the head joint like this, you will not have to return to the mortar pan for each individual head joint. When buttering the head, hold the trowel blade at an angle, so as not to move or cut the line. After practice, disturbing the line can be minimized.

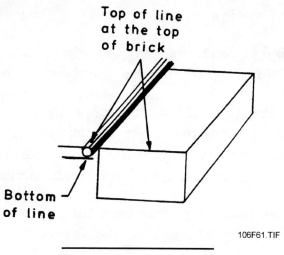

Top of line
at the top
of brick

Bottom
of line

106F61.TIF

Figure 61. Setting Brick

Most of the mason's time is spent laying brick to the line. Practice will improve your ability to lay precisely without disturbing the line constantly. As you learn to do this, there are some additional habits you should pick up to save yourself time and energy.

- Always pick up a brick with the face out so that it is in the same position in which you will lay it. Limit turning brick in your hand, as this slows and tires you.

- Pick up frogged brick with the frog down, since this is the way it will be laid.

- Fill head and bed joints plump and full. This cuts down the time you will need to strike joints later. This also ensures stronger, waterproof walls.

- Stock your brick within arm's length, or approximately 2 feet away. When working on a veneer or cavity wall, stock your brick on the backing wythe.

8.3.3 Laying Block To The Line

The difference in laying block and brick to the line is the difference in handling the units.

Blocks should be kept dry at all times, as moisture will cause them to expand. If they are used wet, they will shrink when they dry and cause cracks in the wall joints. To cut down on handling, stack them close to the work sites with the bottom (smaller) shells and webs down.

Practicing laying block will let you discover the easiest methods for yourself. Find a way to hold the buttered block that is comfortable for you. Lift the block firmly by grabbing the web at each end of it and lay it on the mortar joint. Keep the trowel in your hand when laying block to save time. *Figure 62* shows one way of doing this.

As you place the block, tip it a little toward you. You can look down the face to align the block with the top of the block in the course below. Then, roll the block back slightly so that the top is in correct alignment to the line. At the same time, press the block toward the last block laid. Moving the block slowly is key to this process. Do not release the block quickly or you will have to remortar and reposition it.

Figure 62. Lift Block And Trowel

You can adjust the block by tapping. Be sure to tap in the middle of the block, away from the edges. Block face shells may chip if you tap on them. Using your trowel handle on the block is not recommended, as the block roughens up the end of the handle. Use your mason's hammer.

8.4.0 BUILDING CORNERS AND LEADS

Corners are called leads because they lead the laying of the wall. They set the position, alignment, and elevation of the wall by serving as guides for the courses that fill the space between them. Building corners requires care and accurate leveling and plumbing to assure that the corner is true.

As you learn to build corners, practice technique and good workmanship. Speed will follow. Be certain that each course is properly positioned before going on to the next. Once a corner is out of alignment, it is difficult to straighten it.

In addition to corners being leads, masons also have leads that are not corners. The next sections discuss leads that are and are not corners.

8.4.1 Rackback Leads

Sometimes it is necessary to build a lead or guide between corners on a long wall. This is a lead without corner angles, or **returns**. It is merely a number of brick courses laid to a given point. A **rackback** lead is **racked**, or stepped back a half brick on each end. This means that the lead is laid in a ½-lap running bond with one less brick in each course.

The first course is usually six bricks long, the length of the mason's level. Each course is one brick less, until the sixth course has only a single brick, pyramid fashion. *Figure 63* shows a completed racked lead.

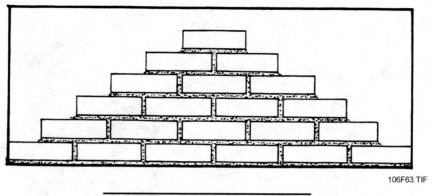

Figure 63. Straight Racked Lead

Building a lead starts with marking the exact place in line with the corners and properly located for the bond pattern. Use a chalk line between the corners to locate the place. Lay brick in the rackback lead by following standard techniques. As each course is laid, check the course spacing on each course with the modular or spacing rule. Then check the course level, plumb, and straightness with your mason's level.

When the lead is complete, it needs to be checked for diagonal alignment. Do this by holding the mason's level at an angle along the side edges of the end bricks. As shown in *Figure 64*, hold the rule in line with the corner of each brick. This lets you check that no bricks are protruding.

Figure 64. Diagonally Aligning The Lead

Then **tail** the diagonal by laying your mason's level on the points of the end bricks, as shown in *Figure 65*. If the rule touches the edges of all the bricks, the head joint spacing is correct. If the head joint spacing is not correct, take out one or two units in that course and the courses above it. Clean and reset them with fresh mortar.

The rackback lead can now be used to anchor a line as detailed previously. Learning to build rackback leads will teach you three-quarters of what you need to know about building corners.

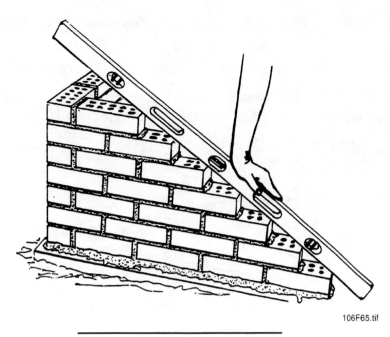

Figure 65. Tailing The Diagonal

8.4.2 Brick Rackback Corners

A rackback corner is a rackback lead with a return or bend in it. The return must be a 90-degree angle, unless the specifications say otherwise. Placement and alignment of the corner are crucial because the corner will set the location of the remainder of the wall. Laying a corner can be intricate, demanding work if there is a pattern bond to follow. The following steps are for building an unreinforced outside rackback corner in a ½-lap running bond pattern.

Step 1 Check the specifications for the location of the corner. Determine how high the corner should be. The corner should reach halfway to the top of a wall built with no scaffold, or halfway to the bottom of the scaffold. Subsequent corners are built as the work progresses.

Step 2 Determine the number of courses the corner will need. Use your course rule to calculate the courses in a given height, then calculate the number of bricks in each course. The sum of the stretchers in the first course must equal the number of courses high the corner will reach. If, for instance, you need a corner 11 courses high, the first course will have 6 bricks on one leg, 5 bricks on the other.

Step 3 Locate the building line and the position of the corner on the footing or foundation. Clean off the footing and check that it is level. Lay out the corner with a steel square. Mark the location directly on the footing. Check the plan or specifications to determine which face of the corner gets the full stretcher and which face gets the header. Some plans have detailed drawings of corners.

Step 4 Dry bond the units along the first course in each leg of the corner. Mark the spacing along the footing.

Step 5   Lay the first course in mortar and check the height, level, plumb, and straightness. Be sure to level the corner brick and the end brick before the bricks in the center. Check height, level, plumb, and straightness for each leg of the corner.

Step 6   Check level on the diagonal, as in *Figure 66*. Lay the mason's level across each diagonal pair of bricks. This will let you make sure that the corner continues level across the angle.

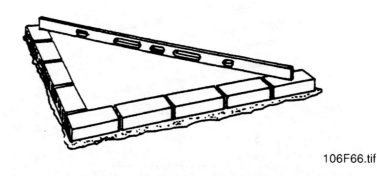

106F66.tif

Figure 66. Leveling The Diagonal

Step 7   Remove excess mortar along the bed and head joints and from the leg ends. Also remove excess mortar from the inside of the corner.

Step 8   Range the bricks. **Ranging** is sighting along a string to check horizontal alignment. Ranging is done after the first course is laid. Fasten one end of the line to the edge of one corner leg and wrap the line around the outside of the corner. Fasten the other end of the line to the outside edge of the other corner leg. Adjust any bricks in the line that are not in perfect horizontal alignment with the line. Then take the line off.

Step 9   Lay the second course, reversing the placement of the bricks at the corner. The leg that had the full stretcher before now gets the header. Use one less brick in the second course to rack the ends. Since each course is racked, stop spreading mortar half a brick from the end of each course.

Step 10   After placing the bricks, check height, level, plumb, and straightness. Check across the diagonal as well. To train your eye, sight down the outermost point of the corner bricks from above to check plumb. Remove excess mortar from the outside and inside of the corner and from each exposed edge brick.

Step 11   Continue until you have reached the required number of courses. Use one less brick in each course. Lay and check each course. Remove excess mortar from each course, inside and outside. Be sure that aligning the bricks does not disturb the mortar bond. If the mortar bond is disturbed, take up the brick, clean it, rebutter it, and replace it.

Step 12   If the corner does not measure up at each course, take the course up and do it again. This is easier than taking the wall up and doing that again.

MASONRY — TRAINEE TASK MODULE 28106

Step 13   When the corner is at the required number of courses, check each leg of the corner for diagonal alignment just as you did for the rackback lead. Hold the mason's level at an angle along the sides of the end bricks, as in *Figure 64*. This lets you check that no bricks are protruding. Also set the mason's level on top of the racked edges, as in *Figure 65*. This lets you check that each one touches the level and that the head joint spacing is correct.

Step 14   Check the mortar and strike the joints. Brush the loose mortar carefully from the brick. Check the height of the corner with the modular spacing rule. Recheck the corner to make sure it is plumb, level, and straight. If it does not measure up, take the corner down and start over.

Because the corner is so important to the wall, speed is not half as important as accuracy. Learn to be accurate, and the speed will follow.

8.4.3   Block Rackback Corners

Speed is the main advantage in building with block. It takes an experienced mason about 40 minutes to lay a block corner compared to 180 minutes to lay a brick corner. Because they are larger and heavier, blocks require some special handling. They chip easily when moved or tapped down in place. They must be eased slowly into position. Their size makes them harder to keep level and plumb. Each block needs to be checked for position in all dimensions. But even with these disadvantages, they do save time and money.

The procedure for laying a block rackback outside corner follows the same steps as for brick. The main difference is that block requires different mortaring and more checking with the mason's level. The steps below do not repeat location material covered above.

Step 1   Clean the foundation and dampen it. Locate the point of the corner. Snap a chalk line from this point across the wall location to the opposite corner. Repeat the procedure on the other side of the corner. This aligns the corner with the other corners. Check the accuracy of the chalk line with a steel square before laying any block.

Step 2   Check that the footing is level. Lay the first course as a dry bond. Because actual sizes of block may vary, space out the dry bond with bits of wood for the joints. Check that the dry bond is plumb to find any irregularities in the footing. If the footing is too high in places, make an adjustment by cutting off some of the bottom of the block. If the footing is too low, add some pieces of block to bring the first course to the correct level.

Step 3   Lay the corner block first. Use a full bedded mortar joint without a furrow. Check the corner block for height, level, plumb, and straightness. Check for plumb on both sides of the block.

Step 4   Continue with the leg of the corner. Line up each block with the chalk line, checking the placement of each block for height, level, plumb, and straightness. Check for alignment as well. When both legs are finished, check them again. Check the diagonals, also.

Step 5   Do not remove excess mortar immediately, as it could cause the block to settle unevenly. Remove excess mortar from the first course after the second course is laid.

Step 6   For subsequent courses, apply mortar in a face shell bedding on top of the previously laid course. Check each block for height, level, plumb, and straightness.

Step 7   As last steps, check for diagonal alignment, then tail the rack ends. If the edges of all blocks do not touch the level, the head joints are not properly sized. Adjust the block if the mortar is still plastic enough or rebuild the courses, as required.

When measuring blocks, each block must touch and be completely flush with the mason's level. They must also be completely in line with the chalk marks on the footing. All measurements must be repeated often to prevent bulges or depressions in the wall and to keep the courses in line.

9.0.0   MORTAR JOINTS

Mortar joints between masonry units do many jobs.

- They bond units together.
- They compensate for differences in size of units.
- They bond metal reinforcements, grids, and anchor bolts.
- They make the structure weathertight.
- They create a neat, uniform appearance.

Mortar joints are made by buttering masonry units with mortar and laying the units. The mason controls the amount of mortar buttered so that it fills a standard space between the units. Excess mortar oozes out between the units, and the mason trims it off. But this is not the last stage in making a mortar joint. After it partially dries, the mortar left between the masonry units must be tooled in some way to be a proper mortar joint.

It is this last step—the tooling—that gives the mortar joints their uniform appearance and weathertight quality.

9.1.0   JOINT FINISHES

Mortar joints can be finished in a number of ways. *Figure 67* shows some standard joint finishes. Usually, the joint finish will be part of the detailed specifications on a project. The process of tooling the joint compresses the mortar and thereby increases its water resistance. Tooling also closes any hairline cracks that open as the mortar dries. Joints are tooled by shaped jointers. Raked joints are made by rakers. Struck, weathered, and flush joints are tooled by a trowel.

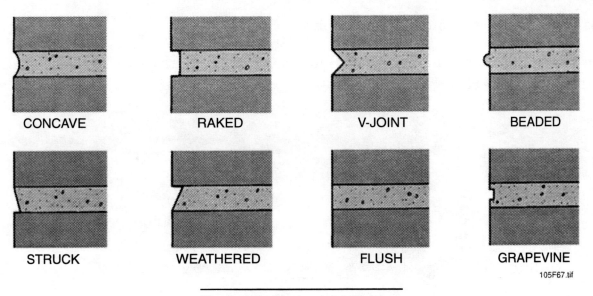

CONCAVE　　RAKED　　V-JOINT　　BEADED

STRUCK　　WEATHERED　　FLUSH　　GRAPEVINE

105F67.tif

Figure 67. Mortar Joint Finishes

The raked joint is scooped out, not compressed. It does not get the extra water resistance, so it is not recommended for exterior walls in wet climates. The struck joint collects dirt and water on the ledge, so it is not recommended for exterior walls. Flush joints are not compressed, only struck off. They are recommended for walls that will be plastered or parged.

One additional joint type is the extruded or weeping joint, *Figure 68*. This joint is made when the masonry unit is laid. When the unit is placed, the excess mortar is not trimmed off. The mortar is left to harden to become the extruded joint. Since the mortar is not compressed in any way, this joint is not recommended for exterior walls.

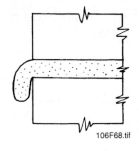

106F68.tif

Figure 68. Extruded Or Weeping Joint

9.2.0 STRIKING THE JOINTS

Working the joints with the jointer, raker, or trowel is called striking the joints. Whichever tool you will use to strike the joint, the procedure will be the same. The first step is to test the mortar.

9.2.1 Testing The Mortar

After the masonry units have been laid, the mortar must dry out before it can be tooled. The test equipment for checking proper dryness is the mason's thumb.

Press your thumb firmly into the mortar joint.

- If your thumb makes an impression, but mortar does not stick to it, the joint is ready, as shown in *Figure 69*.

Figure 69. Thumbprint Test For Jointing

- If the mortar sticks to your thumb, it is too soft. The mortar is still runny, and the joint will not hold the imprint of the jointing tool.
- If your thumb does not make an impression, the mortar is too stiff. Working the steel tool will burn black marks on the joints.

The best time for striking the joints will vary because weather affects mortar drying time. Test the mortar repeatedly to find the right window of time to do the finishing work.

9.2.2 Striking

When the mortar is ready, the next step is the striking. The convex sled runner striking tool is most commonly used today. The tool should be slightly larger than the mortar joint to get the proper impression.

- Hold the tool with your thumb on the handle so it does not scrape on the masonry unit.
- Apply enough pressure so that the runner fits snugly against the edges of the masonry units. Keep the runner pressed against the unit edges all the way through the strike.
- Strike the head joints first, as in *Figure 70*. Strike head joints upward for a cleaner finish.

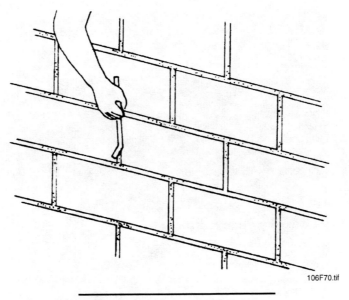

Figure 70. Striking Head Joints

- Strike the bed joints last. To keep them smooth, walk the jointer along the wall as you strike. The joints should be straight and unbroken from one end of the wall to the other. If the head joints are struck last, they will leave ridges on the bed joints.

If you are making a raked joint, follow the same order of work. Some joint rakers, or skate rakers, have adjustable set screws that set the depth of the rake-out. Be sure you do not rake out more than ½ an inch or you will weaken the joint and possibly expose ties or reinforcements in the joint. Be sure you leave no mortar on the ledge of the raked unit.

If you are making a troweled joint, follow the same order of work. Be sure the angle of the struck or weathered joint faces the same way on all the head joints. If you are striking flush joints with your trowel, strike up rather than down.

9.2.3   Cleaning Up Excess Mortar

After you strike the joints, you must clean up the excess mortar. Dried mortar sticks to masonry and is difficult to clean. Cleaning immediately after striking is the easier path to follow.

- Trim off mortar burrs by using a trowel. Hold the trowel fairly flat to the wall, as shown in *Figure 71*. As you trim the burrs, flick them away so they do not stick to the units.
- Dress the wall after trimming off burrs. Dressing can be done with a soft brush or coarse fabric such as burlap or carpet wrapped around a wood block. Flush joints need the fabric dressing, as a brush will not smooth them.
- Brush the head joints vertically first. Then brush the bed joints horizontally. If necessary, restrike the joints after brushing to get a sharp, neat joint.

After you finish cleaning the wall, clean the floor at the foot of the wall.

Figure 71. Trim Mortar Burrs

10.0.0 PATCHING MORTAR

Pointing is the act of putting additional mortar into a soft mortar joint. This type of patch does not require much preliminary work. Tuckpointing is the act of replacing hardened mortar with fresh mortar. This type of patch requires some preparation.

10.1.0 POINTING

Despite the best workmanship, mortar can fall out of a head joint or crack when a unit settles. A unit may get a chipped edge or a lost corner or get moved by some accident. Line pins and nails leave holes in mortar joints. Since pointing is easier than patching, it is a good idea to continuously check the surface of mortar joints. Perform the following tasks as you work:

- Fill line pin and nail holes as you move the line.

- Use mortar of the same consistency as was used for laying the units.

- Force the mortar into the holes with the tip of a pointing trowel or a slicker.

- Push all the mortar with a forward motion, in one direction for each hole.

- If the hole is deep, fill it with several thin layers of mortar, each no more than ¼ of an inch deep. This will avoid air pockets in the pointed joint.

- Clean excess mortar off masonry units with your trowel.

Inspect the condition of mortar joints after you finish a section of wall. Inspect them again before you strike them with the jointer. If there is a void, force additional mortar into the joint. If the back of the unit can be reached, use a backstop, such as the handle of a hammer, to brace the unit. This will prevent the unit from moving as you point the joint.

10.2.0 TUCKPOINTING

Patching or tuckpointing after the mortar has hardened is more complex. Follow these steps for a good job:

Step 1 Mix some mortar and let it dry out for about an hour to get partly stiff. This will reduce shrinkage after it is put into the joint. While it is stiffening, clean out the damaged joints.

Step 2 With a joint chisel or a tuckpointer's grinder, dig out the bad mortar to a depth of about ½ an inch. The damaged area may be deeper due to cracks or shrinkage. Be sure you have cleaned down to solid mortar. As a rule, the depth of mortar removed should be at least as deep as the joint is wide. *Figure 72* shows a properly excavated joint.

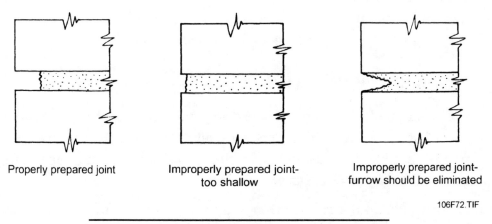

Properly prepared joint

Improperly prepared joint-
too shallow

Improperly prepared joint-
furrow should be eliminated

106F72.TIF

Figure 72. Properly Cut Joints For Tuckpointing

Step 3 Remove all lose mortar with a stiff brush or with a jet of water from a hose.

Step 4 Thoroughly wet the surrounding masonry with water, but do not saturate it. Wetting will slow setting time and produce a better bond. However, excess moisture will bead up and prevent a good bond in the joint.

Step 5 Force fresh mortar into the damp joint. Use a trowel or slicker with a point narrower than the joint and press the mortar hard. If the damaged area is deep, fill it with several thin layers of mortar, each no more than ¼ of an inch deep. This will avoid air pockets in the pointed joint. *Figure 73* shows what happens when a tuckpointed joint is not filled with carefully applied layers.

Step 6 Clean excess mortar off units with your trowel.

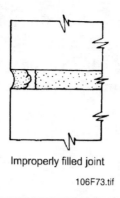

Improperly filled joint

106F73.tif

Figure 73. Improperly Filled Joint

Step 7 Retool the joint after the mortar has set long enough.

Step 8 Clean the pointed areas after you retool the joints. Use the cleaning procedure described previously under *Striking the Joints*.

11.0.0 CLEANING MASONRY UNITS

Cleaning masonry units marks the end of a particular project. The finishing touch on all masonry work is the removal of any dirt and stains. This can be a wet, difficult task. The best way to minimize this effort is by cleaning as the project goes along.

11.1.0 CLEAN MASONRY CHECKLIST

The hardest soil to clean off masonry units is dried, smeared mortar that has worked its way into the surface of the masonry unit. This seriously affects the appearance of the finished structure. Your best approach is to avoid smearing and dropping mortar during construction. These guidelines will help you clean as you work:

• When mortar drops, do not rub it in. Trying to remove wet mortar causes smears. Let it dry to a mostly hardened state.

• Remove it with a trowel, putty knife, or chisel. Try to work the point under the mortar drop and flick it off the masonry.

• The remaining spots can usually be removed by rubbing them with a piece of broken brick or block, then with a stiff brush.

You should spend some time cleaning every day, removing stray mortar from the wall sections as you complete them.

In addition to cleaning dropped mortar there are other things to do. To keep masonry clean during construction, practice these good work habits:

• Stock mortar pans and boards a minimum of 2 feet away from the wall to avoid splashes.

• Temper the mortar with small amounts of water so it will not drip or smear on the units.

• After laying units, cut off excess mortar carefully with the trowel.

- Wait until mortar hardens for striking to avoid smearing wet mortar on masonry units.

- After tooling joints, scrape off mortar burrs with your trowel before brushing.

- Avoid any motion that rubs or presses wet mortar into the face of the masonry unit.

- Keep materials clean, covered, and stored out of the way of concrete, tar, and other staining agents. Do not store materials under the scaffolding.

- Turn scaffolding boards on edge with the clean side to the wall at the end of the day. This will prevent rain from splashing dirt and mortar on the wall.

- Always cover walls at the end of the day to keep them dry and clean.

Following these practices should reduce the amount of time you spend cleaning the masonry units after construction is complete. The clean masonry checklist is especially important when working with CMUs. Because of their rougher surface texture, mortar spilled on them is harder to clean.

11.2.0 BUCKET CLEANING FOR BRICK

The best method of cleaning any new brick masonry is the least severe method. If the daily cleaning practices listed above are not enough, the next step is bucket and brush hand cleaning. This may include using a proprietary cleaning compound or an acid wash. Acid affects brick over time, so it should be considered the cleaner of last resort. *Figure 74* lists cleaning methods for different types of brick as developed by the Brick Institute of America.

Any chemical compound used should first be tested on a 4 × 5 foot inconspicuous section of wall. Sometimes, minerals in the brick may react with some chemicals and cause stains. Read the brick manufacturer's materials safety data sheet for recommended cleaning solutions. Read the materials safety data sheet and follow the manufacturer's directions for mixing, using, and storing any chemical solution.

WARNING!   Wear Appropriate Personal Protective Equipment when using chemical solutions.

When cleaning, follow these guidelines. You will need a hose, bucket, wooden scraper, chisel, and stiff brush.

- Do not start cleaning until at least 1 week after the wall is finished. This gives the mortar time to cure and set. Do not wait longer than 6 months because the mortar will be almost impossible to remove.

- Dry scrub the wall with a wooden paddle. Go over large particles with a chisel, wood scraper, or piece of brick or block. This dry rubbing should remove most of the mortar.

- Before wetting, protect any metal, glass, wood, limestone, and cast stone surfaces. Mask or cover windows, doors, and fancy trim work.

Brick Category	Cleaning Method	Remarks
Red and Red Flashed	Bucket and Brush Hand Cleaning High Pressure Water Sandblasting	Hydrochloric acid solutions, proprietary compounds, and emulsifying agents may be used. *Smooth Texture:* Mortar stains and smears are generearaly easier to remove; less surface area exposed; easier to presoak and rinse; unbroken surface, thus more likely to display poor rinsing, acid staining, poor removal of mortar smears. *Rough Texture:* Mortar and dirt tend to penetrate deep into textures; additional area for water and acid absorption; essential to use pressurized water during rinsing.
Red, Heavy Sand Finish	Bucket and Brush Hand Cleaning High Pressure Water	Clean with plain water and scrub brush, or *lightly* applied high pressure and plain water. Excessive mortar stains may require use of cleaning solutions. *Sandblasting is not recommended.*
Light Colored Units, White, Tan, Buff, Gray, Specks, Pink, Brown and Black	Bucket and Brush Hand Cleaning High Pressure Water Sandblasting	*Do not use muriatic acid!!* Clean with plain water, detergents, emulsifying agents, or suitable proprietary compounds. Manganese colored brick units tend to react to muriatic acid solutions and stain. Light colored brick are more susceptible to "acid burn" and stains, compared to darker units.
Same as Light Colored Units, etc., plus Sand Finish	Bucket and Brush Hand Cleaning High Pressure Water	Lightly apply either method. (See notes for light colored units, etc.) *Sandblasting is not recommended.*
Glazed Brick	Bucket and Brush Hand Cleaning	Wipe glazed surface with soft cloth within a few minutes of laying units. Use soft sponge or brush plus ample water supply for final washing. Use detergents where necessary and acid solutions only for *very difficult mortar* stain. For dilution rate, see Step 1d, *Select the Proper Solution,* under Bucket and Brush Hand Washing. Do not use acid on salt glazed or metallic glazed brick. Do not use abrasive powders.
Colored Mortars	Method is generally controlled by the brick unit	Many manufacturers of colored mortars do not recommend chemical cleaning solutions. Most acids tend to bleach colored mortars. Mild detergent solutions are generally recommended.

106F74.tif

Figure 74. Cleaning Guide For New Masonry

- Prepare the chemical cleaning solution. Follow manufacturer's directions. Remember to pour chemicals into water, not water into chemicals.

- Presoak the wall with the hose to remove loose particles or dirt.

- Start working from the top. Keep the area immediately below the space you are scrubbing wet also to prevent the chemical from drying into the wall.

- Scrub a small area with the chemical applied on a stiff brush. Keep the scrub area small enough so that the solution does not dry on the wall as you are working.

- To remove stubborn spots, rub a piece of brick over them. Then rescrub the spot with more chemical solution. Repeat this as needed.

- As you complete scrubbing each area, rinse the wall thoroughly. Rinse the surrounding wall area above and below, all the way to the bottom of the wall, to keep chemicals from staining the wall.

- Flush the entire wall for 10 minutes after completing scrubbing. This will dilute any remaining chemical and prevent burns.

High pressure water washing, steam cleaning, and sandblasting are also used to clean new and old masonry. Because these techniques can damage masonry surfaces, they require trained operators. If you have not been trained, do not use this equipment.

11.3.0 CLEANING BLOCK

Cleaning block is difficult, as the surface on standard block is very porous. It is important to follow the *Clean Masonry Checklist* procedures above. If block is stained with mortar at the end of a job, rub it with a piece of block.

For further cleaning, it is important to check the manufacturer's materials safety data sheets for recommended chemical cleaners and cleaning procedures. Acid is very destructive of block and cannot be used without protective countermeasures. Read the block manufacturer's materials safety data sheets for recommended cleaning solutions. Read the materials safety data sheets and follow manufacturer's directions for mixing, using, and storing any chemical solution. Detergents and surfactants are often recommended for use on block. If no cleaning procedures are given, follow the bucket and brush procedures listed above.

Any chemical compound used should first be tested on a 4 × 5 foot inconspicuous section of wall. Sometimes, minerals in the block may react with some chemicals and cause stains.

In addition to mortar, block is easily stained by many other substances. *Figure 75* presents a chart that contains a list of staining agents and recommended cleaners that can be used under most circumstances. The last column in *Figure 75* refers to *Figure 76*. This chart contains procedures for applying the cleaners detailed in *Figure 75*. If you are trying to clean a stain on a block wall, first look in *Figure 75* to identify the type of stain. Then look up the procedure references in *Figure 76* and follow the referenced procedures.

Stain	Appearance	Materials Needed		Procedural Sequence (Letters refer to steps listed in Figure 29-8)
		Chemicals and Detergents*	Poultice Materials	
Aluminum	White deposit	10% Hydrochloric acid	—	A-B
Asphalt	Black	1.* Ice. (Dry ice is not very effective)	—	C
		2.* Scouring powder	—	A
Emulsified Asphalt	Black	Water	—	A₁
Cutback Asphalt	Black	1.* Benzene	Talc or whiting	D₁-D₂-D₃
		2.* Scouring powder	—	A
		3.* Same	Same	E
Coffee	Tan	Sodium hypochlorite OR Glycerine 1 part Water 4 parts	—	F
Copper, bronze	Green, sometimes brown	Ammonium chloride 1 part Ammonium hydroxide as needed for paste	Talc 4 parts	D-D₁-D₂-D₅-E
Creosote	Brown	1.* Benzene	Talc or whiting	D₁-D₂-D₅
		2.* Scouring powder		A
Ink, ordinary blue	Blue	Sodium perborate Water	Whiting	G-D₁-D₂-D₅-E-E₁
Containing Prussian Blue	Blue	Ammonium hydroxide OR Strong soap solution	—	F
Red, green, violet, other bright colors and indelible synthetic dyes	Varies	Sodium perborate Water OR Sodium hypochlorite OR Calcium hypochlorite	Whiting	G-D₁-D₂-D₅-E
		OR Ammonium hydroxide OR Sodium hypochlorite OR Calcium hypochlorite	—	F
Indelible, containing silver salt	Black	Ammonium hydroxide	—	F-E
Iodine	Brown	Ethyl alcohol	Whiting or talc	H₁-D₁-D₂-D₅
		OR None	—	J
Iron	Brown yellow	Sodium or ammonium citrate 1 part Water (lukewarm) 6 parts Glycerine (lime-free) 7 parts	Whiting or Diatomaceous earth	G₁-K-D₁-D₂-D₅-E-B
		OR Same plus sodium hydrosulfite for step L	Same	G₁-K-D₁-D₂-L-D₅-B
Linseed oil	Dark gray	Trisodium phosphate 1 part Sodium perborate 1 part Liquid green soap or strong soap solution in hot water	Lime, whiting, talc or portland cement	M-G₁-D₁-D₂-D₅-B-E
Lubricating oil or grease that has penetrated	Dark	1.* Trisodium phosphate 1 lb Water 1 gal.	—	N-G₁-A
		2.* Benzene	Talc, lime or whiting	D₁-D₂-D₅-B
		OR Amyl acetate (for small areas only)	Asbestos fiber	D₁-D₂-D₃-D₅
Mortar smears	Gray	—	—	O
Paint, at least 3 days old	Varies	Trisodium phosphate 2 lb. Water 1 gal OR Commercial paint remover	—	H-P
freshly spilled		Same		S-R-H-P
Perspiration stains from hands, and hair oil stains	Brown or yellow	Trichloroethylene	Talc	T-D₁-D₂-D₄-D₅
Plant growth, mold and moss	Green, brown or black	Ammonium sulfamate (from garden supply stores)	—	V, and B if needed
Smoke and fire	Brown to black	Trichloroethylene	Talc	T-D₁-D₂-D₄-D₅
Soot and coal smoke	Black	Soap Water Pumice	—	A
Wood tar and smoke	Dense black	1.* Scouring powder Water	—	A
		2.* Sodium hypochlorite	—	F
Tobacco	Brown	1.* Calcium chloride Water	—	D₁
		2.* Trisodium phosphate 2 lb Water 5 qt.	—	G₁-X
		3.* —	Talc	D₁-D₂-D₅-E-B
Wood rot	Chocolate	1.* Glycerine 1 part Water 4 parts	—	A
		2.* Trichloroethylene	Talc	T-D₁-D₂-D₄-D₅

*Numbers indicate that materials are to be used in sequence.

106F75 tif

Figure 75. Stains And Stain Removers

PROCEDURE FOR STAIN REMOVAL

A	Scrub with brush and chemicals or detergents
A_1	Scrub with brush and water
B	Wash thoroughly with clear water
C	Cool until brittle. Chip away with chisel
D	Mix solids
D_1	Stir solids and liquid to thick paste
D_2	Apply paste to stain to thickness of 1/8 to 1/4 inch
D_3	Place heated concrete brick over top to drive liquid in
D_4	Cover to minimize rate of evaporation
D_5	Let dry as needed for periods up to 24 hours. Scrape off. Use wood scraper if block has tile-like finish
E	Repeat as needed
E_1	If brown stain remains, treat as for iron stain
F	Apply with saturated cloth or cotton batting
G	Dissolve solid chemical in hot water
G_1	Dissolve solid chemicals in water

H	Apply liquid to surface by brush
H_1	Apply liquid to surface by brush at 5 to 10 minute intervals until thoroughly soaked
J	Allow stain to disappear by aging
K	Stir liquids together
L	Put paste on trowel. Sprinkle crystals on top of paste. Apply to surface so crystals are in contact with block and paste is on outside.
M	Soak up fast and soon with poultice material until no free oil remains
N	Scrap any solidified matter off surface
O	Let harden. Remove large particles with trowel, putty knife or chisel
P	Let stand. Remove with scraper and wire brush.
R	Allow to age three days
S	Absorb with soft cloth or paper towels, then scrub vigorously with paper towels
T	Provide thorough ventilation
V	Follow manufacturer's directions
X	Pour into paste and mix

106F76.TIF

Figure 76. Procedure For Stain Removal

SUMMARY

This module has described common masonry units and accessories. It contains detailed instructions for locating and laying out walls and corners. It also gives procedures for laying a dry bond. It details spreading mortar on units and laying them. It differentiates pattern and structural bonds and shows common structural pattern bonds. It provides instruction in cutting brick and block with chisels or masonry hammers, saws, and splitters. The heart of this module is concerned with laying masonry units in walls, rackback leads, and rackback corners. It further discusses striking joints, pointing and tuckpointing, and cleaning masonry units.

References

For advanced study of topics covered in this task module, the following reference materials are suggested:

Building Block Walls—A Basic Guide, National Concrete Masonry Association, Herndon, VA, 1988.

Masonry Design and Detailing—For Architects, Engineers and Contractors, Fourth Edition, Christine Beall, McGraw-Hill, New York, NY, 1997.

Masonry Skills, Third Edition, Richard T. Kreh, Delmar Publishers, Inc., Albany, NY, 1990.

The ABCs of Concrete Masonry Construction, Videotape 13:34 minutes, Portland Cement Association, Skokie, IL, 1980.

ACKNOWLEDGMENTS

Figures 1, 10-11, 20, 68, and 71-73 courtesy of The Portland Cement Association
Figures 2 and 42 courtesy of Carolinas Concrete Masonry Association
Figures 3, 13-19, 22, 24-26, 34, and 70 courtesy of National Concrete Masonry Association
Figures 21, 29-30, 32-33, 36, 46-51, 53-54, 56-58, 60, 62, 69, and 74-76 courtesy of Delmar Publishers, Inc.
Figures 4, 7-9, 23, and 43-44 courtesy of The Goodheart-Willcox Company, Inc.
Figures 12, 31, 35, and 37-41 courtesy of McGraw-Hill
Figures 28 and 59 courtesy of Roy Jorgensen Associates, Inc.
Figure 5 courtesy of United Concrete Products, Frederick, MD

REVIEW/PRACTICE QUESTIONS

1. Random cracking in CMU walls can be controlled by _____.

   a. contraction joints
   b. changes in wall thickness
   c. pilasters
   d. chases

2. Aggregates, portland cement, admixtures, and water make up _____.

   a. mortar
   b. hydrated lime
   c. concrete blocks
   d. veneer ties

3. Hollow masonry units have _____.

   a. 75 percent or more of their surface solid
   b. 75 percent or more of their surface hollow
   c. 50 percent or more of their surface solid
   d. 25 percent or less of their surface hollow

4. Calcium silicate brick is unusual in that it is _____.

   a. used mostly in Europe
   b. not fired in a kiln
   c. good for fireplaces and chimneys
   d. made without cement and sand

5. Cavity walls can be loadbearing when they _____.

   a. have flashing and expansion joints in the cavity
   b. are placed on a modular grid to avoid cuts
   c. are tied with metal ties to equalize bearing forces
   d. have both wythes of the same material

6. Laying the first course with no mortar is called _____.

   a. leading
   b. rackbacking
   c. hot ranging
   d. dry bonding

7. Picking up mortar requires _____ with the trowel.

   a. rotating and spreading
   b. scooping and shaving
   c. tilting and cutting
   d. cutting and scooping

8. Use a small snap of your wrist to _____.

    a. set mortar firmly on the trowel blade

    b. release mortar from the trowel blade

    c. set mortar firmly on the bed joint

    d. set mortar firmly on the head joint

9. Stringing is _____.

    a. mortaring the ends of the masonry units

    b. keeping the mortar from drying out

    c. part of a pilaster

    d. spreading a bed joint with mortar

10. Putting a trough in the bed joint is _____.

    a. furrowing

    b. rackbacking

    c. tailing

    d. tuckpointing

11. Moving a CMU after it is in place will _____.

    a. chip its edges

    b. weaken the mechanical bond between the mortar and the unit

    c. set mortar firmly

    d. retemper the mortar

12. A pattern bond does not _____.

    a. call for cutting brick to fit

    b. add a visual design to masonry walls

    c. add structural strength to masonry walls

    d. call for placing brick in different positions

13. Flemish, common, and running are _____.

    a. stack bonds

    b. defined brick cuts

    c. corner patterns

    d. structural pattern bonds

14. Cutting brick or block allows for _____.

    a. removing burrs

    b. filling around openings

    c. leveling courses

    d. standardizing shapes and sizes

15. Cutting masonry with a splitter calls for _____.

    a.   venting the dust

    b.   using a diamond blade

    c.   checking its condition before using it

    d.   using a pusher or conveyor

16. Plumbing and leveling masonry units is not done _____.

    a.   for every six bricks and every single block

    b.   before cutting off extruded mortar

    c.   by tapping units into place

    d.   after striking joints

17. Laying to the line calls for positioning the masonry unit _____.

    a.   level with the bottom of the string and $\frac{1}{16}$ of an inch away

    b.   level with the top of the string and $\frac{1}{16}$ of an inch away

    c.   level with the bottom of the string and $\frac{3}{32}$ of an inch away

    d.   level with the top of the string and $\frac{3}{16}$ of an inch away

18. Poles, blocks, pins, and trigs are used to _____.

    a.   level masonry units

    b.   strike mortar joints

    c.   secure mason's line

    d.   shield masonry from mortar smears

19. Laying to the line is harder when you _____.

    a.   stock materials close by

    b.   handle and turn units as little as possible

    c.   use trimmed mortar to butter the next head joint

    d.   frequently crowd the line

20. A rackback lead is _____.

    a.   stepped back a half brick on each side

    b.   used to check plumb

    c.   slack to the line

    d.   crowding the line

21. A lead with a return in it is a _____.

    a.   furrow

    b.   corner

    c.   string

    d.   rackback

22. Joints should be struck when the mortar _____.
    a. brushes off the masonry units
    b. has set firm and hard
    c. will hold a thumbprint
    d. still oozes

23. The difference between tuckpointing and pointing is _____.
    a. striking the joints
    b. working after or before the mortar finishes hardening
    c. trimming the burrs
    d. scooping the mortar

24. Masons clean up more _____ than anything else.
    a. mortar
    b. sand
    c. admixtures
    d. efflorescence

25. Cleaning with acid is not _____.
    a. recommended for brick
    b. recommended for block
    c. potentially dangerous for the cleaner
    d. damaging to masonry surfaces

notes

ANSWERS TO REVIEW/PRACTICE QUESTIONS

<u>Answers</u>		<u>Section Reference</u>
1.	a	2.2.0
2.	c	2.3.0
3.	b	2.4.0
4.	b	2.6.2
5.	c	3.3.1
6.	d	4.2.3
7.	d	5.1.1, 5.1.2
8.	a	5.1.2
9.	d	5.2.1
10.	a	5.2.3
11.	b	5.3.2
12.	c	6.2.0
13.	d	6.3.0
14.	b	7.0.0
15.	c	7.3.1
16.	d	8.1.1
17.	b	8.3.0
18.	c	8.3.1
19.	d	8.3.2
20.	a	8.4.1
21.	b	8.4.2
22.	c	9.2.1
23.	b	10.0.0
24.	a	11.1.0
25.	b	11.3.0

The NCCER makes every effort to keep these manuals up-to-date and free of technical errors. We appreciate your help in this process. If you have an idea for improving this manual, or if you find an error, a typographical mistake, or an inaccuracy in the NCCER's Craft Training Manuals, please write us, using this form or a photocopy. Be sure to include the exact module number, page number, a description of the problem, and the correction, if possible. Your input will be brought to the attention of the Technical Review Committee. Thank you for your assistance.

Instructors – If you found that additional materials were necessary in order to teach this module effectively, please let us know so that we may include them in the Equipment/Materials list in the Instructor's Guide.

Write: Curriculum and Revision Department
National Center for Construction Education and Research
P.O. Box 141104
Gainesville, FL 32614-1104
Fax: 352-334-0932

Craft _____ Module Name _____

Module Number _____ Page Number(s) _____

Description of Problem _____

_____

_____

_____

_____

(Optional) Correction of Problem _____

_____

_____

(Optional) Your Name and Address _____

_____

_____

_____